Alcuin's Recreational Mathematics

Alcuin's Recreational Mathematics

River Crossings and other Timeless Puzzles

MARCEL DANESI

OXFORD
UNIVERSITY PRESS

OXFORD
UNIVERSITY PRESS

Great Clarendon Street, Oxford, OX2 6DP,
United Kingdom

Oxford University Press is a department of the University of Oxford.
It furthers the University's objective of excellence in research, scholarship,
and education by publishing worldwide. Oxford is a registered trademark of
Oxford University Press in the UK and in certain other countries

Published in the United States of America by Oxford University Press
198 Madison Avenue, New York, NY 10016, United States of America

British Library Cataloguing in Publication Data
Data available

Library of Congress Control Number is on file with the LOC.

ISBN 9780198925309

DOI: 10.1093/9780198925330.001.0001

Printed and bound by
CPI Group (UK) Ltd, Croydon, CR0 4YY

Contents

Acknowledgments

I have discussed most of the ideas in this book with my students in a course I taught for several decades at Victoria College at the University of Toronto called "Puzzles, Discovery, and the Human Imagination." Any infelicities that this book contains are my sole responsibility. I sincerely hope that students, teachers, mathematicians, and general readers alike will glean something interesting from it. As the great mathematician, David Hilbert, once observed, "Mathematics is a game played according to certain rules with meaningless marks on paper" (cited in Pillis and Rose 1988). The creative source of those marks is often to be found in math puzzles themselves.

Introduction

At some time or other, you may have come across this simple, yet still thought-provoking, puzzle:

> A traveler comes to a riverbank with a wolf, a goat, and a head of cabbage. He sees a boat there that he can use for crossing over to the other bank; but to his dismay, he notices that it can carry no more than two—the traveler himself and just one other. As the traveler knows, if left alone together on either bank, the goat will eat the cabbage and the wolf will eat the goat. The wolf does not eat cabbage. How does the traveler transport the animals and the cabbage to the other side intact in a minimum number of trips?

While the puzzle might seem to be trivial mathematically, it actually enfolds a way of thinking that puts on display what mathematics is all about in a nutshell. It also enfolds ideas that have come to fruition in mathematics and computer science. Its solution and implications will be discussed in detail in Chapter 7 of this book. Known aptly as a "river-crossing problem," it has been translated into many languages since it appeared in a little book of problems titled *Propositiones ad acuendos juvenes* ("Problems to Sharpen the Young"), compiled around 800 CE by the Northumbrian theologian and educator, Alcuin of York (735–804 CE), who went officially by the Latin name of Flaccus Albinus Alcuinus. Alcuin was called in 782 by the Emperor Charlemagne to his court with the task of helping him facilitate the spread of education throughout his empire, which came eventually to be called the "Carolingian Renaissance."

From most accounts, Alcuin composed the *Propositiones* for young students during his tenure at the court, designing it as a problem-solving manual in mathematics, seemingly guided by the view that by "doing" intriguing and challenging problems, the "learning" of mathematics would be rendered more enjoyable and thus greatly facilitated—a pedagogical concept that has remained a firm one in math education to this day.

The *Propositiones*

There is some debate as to whether Alcuin was the actual writer of the *Propositiones*, with some historians attributing its content to Boethius (c. 480–524), the Roman senator and scholar, and others maintaining that it may have been the work of the Venerable Bede (672–735 CE), the English monk, theologian, and historian. However, as David Singmaster (1998) has cogently argued, there is no solid historical evidence to suggest that Alcuin copied or adapted his work from previous authors or sources, even though a few of his problems appear in different languages and time eras, as Martha Ascher (1991) has documented. Since it is unlikely that Alcuin knew the different languages, the similarity of some of his most ingenious problems to them would imply that they reflect universal ideas that find expression in mathematics, independently of language or culture. A few of Alcuin's problems do, however, seem to have their basis in the *Greek Anthology*, written a few centuries earlier.

The Greek Anthology

Spanning the Classical and Byzantine periods of Greek civilization, this is a compilation of literary verses, epigrams, riddles, and mathematical puzzles, known far and wide and likely used for the education of children. Some credit the poet Metrodorus (c. 500) with writing it, although the actual author is not known for certain. The *Anthology* is interesting historically because it recycles ancient mathematical problems, spreading them throughout the medieval period.

Written centuries before the advent of mass printing technologies, Alcuin's book was copied by numerous scribes, which is the most plausible reason why different editions of the *Propositiones* present variations that are due to copying errors or unwarranted attempts to "improve" the text by certain scribes. One major difference is the number of puzzles, with some versions containing 53 and others 56.

Alcuin's book has all the hallmarks of a math textbook designed for students. He introduces each problem with a thematic heading, followed by a statement of the problem, and then a formulaic request for an answer. Finally, Alcuin provides a solution, without explaining it fully, which he probably left up to the teachers who used the text to do so. A few of the problems, such as those based on tricky logic (Chapter 6 in this book), could have been easily deciphered by students with no recourse to previous mathematics; but most required some mathematical training and would have taken quite some effort to solve, especially because of the cumbersome Roman numeral system in place at the time. However, no problem in the text requires a sophisticated technical knowledge—just elementary mathematical thinking (which was likely the main objective of the text).

There have been several English editions of the *Propositiones*. The three most widely used ones are: (1) by John Hadley and David Singmaster in 1992; (2) by Peter J. Burkholder in 1993; (3) and the online one by J. J. O'Connor and E. F. Robertson in 2012. All provide an update of the medieval language of Alcuin's times as well as commentaries and annotations to the problems. While I have consulted these translations, I have provided my own slightly different versions, attempting to make them as accessible as possible to people today.

Recreational Mathematics

The problems in the *Propositiones* are *de facto* puzzles in the contemporary sense of the word. For this reason, the book is often pegged as a foundational one in *recreational mathematics*, as we know it today. To cite David Singmaster (1998: 11): "The *Propositiones* of Alcuin is a major landmark in the history of mathematics in general as well as in the history of recreational mathematics in particular."

Puzzles, Propositions, Problems

These three terms will come up throughout this book, and will be used somewhat interchangeably, for the sake of convenience, although there are differences. The English term *puzzle*, to describe a tricky problem, did not exist in antiquity or the medieval period. It emerged in the English-speaking world in the Renaissance period, becoming a general moniker after the invention of the jigsaw puzzle around 1760. The term *proposition* was introduced by ancient mathematicians in reference to a statement whose veracity had to be proved logically. The word *problem* was coined at around the same time in reference to any mathematical proposition that led a specific solution. In the medieval period, it was not uncommon to use *proposition* and *problem* as synonyms—hence, Alcuin's use of *Propositiones* for the title of his book.

The *Propositiones* became popular broadly, transcending its pedagogical uses, corroborating the designation of the book as one in recreational mathematics. Indeed, most of the problems are, as mentioned, puzzles in the modern sense of that word—that is, "brain teasers" whose answers are hardly obvious at first, requiring a large dose of imaginative thinking to reach. It is believed that Charlemagne himself enjoyed doing the puzzles, and that Alcuin may have even designed the river crossing puzzle (above) specifically for the emperor. The popularity of the *Propositiones* was matched by several other medieval compilations. One of these was a collection of a hundred mechanical problems titled *Kitab al-Hiyal* ("The Book of Ingenious Devices"), published in 850 CE by three brothers, Ahmad, Muhammad, and Hasan ibn Musa ibn Shakir in Baghdad. But undoubtedly the best-known one comes four centuries after Alcuin, namely, Leonardo Fibonacci's *Liber Abaci*, published in 1202, which will be discussed subsequently in this book.

Purpose of This Book

Alcuin's text emphasizes that mathematics is hardly a boring activity, consisting of adding, subtracting, multiplying, and dividing numbers. It requires the use of ingenious thinking that everyone possesses, but which needs to be fleshed out and practiced methodically. The underlying message of Alcuin's book is, thus, that familiarity with this form of thinking is critical to a well-rounded education, since it stimulates the imagination. The objective of the present book is to unravel the thought processes involved in solving the problems-puzzles in the *Propositiones*, extending them to how mathematics is carried out more broadly. I have written it for a general audience, who may not be aware of Alcuin's text and its historical importance. I have taken nothing for granted, other than a few high school notions. I have used separate boxes throughout the book to fill in what may be potential gaps in knowledge.

Each of the 10 chapters is organized around a set of Alcuin's problems classified under a specific theme. Solutions to the problems follow and are explained in detail. The chapter then annotates and illustrates notions or themes that are inherent in the problems, ending with 10 puzzle explorations, with the answers and detailed solutions provided at the back of the book. I will use algebraic notation throughout, even though this was not employed by Alcuin, for the sake of convenience. The idea of representing numbers with symbols is an ancient one, going back to the Greek mathematician Diophantus (c. 200–284), who will be discussed in Chapter 5. But it was the Persian polymath, Muhammad al-Khwārizmī, who introduced the techniques that we use to this day in algebra.

Muhammad ibn Mūsā al-Khwārizmī (c. 780–850)

The word *algebra* comes from the Arabic word *al-jabr*, meaning "restoration" or "reunion," which is part of the title of one of the most important books in the history of mathematics, Muhammad ibn Mūsā al-Khwārizmī's, *Calculation by Restoration and Reduction* (in Arabic, *Hisab al-jabr w'al-muqabala*). Incidentally, the term *algorithm* is derived from his name. Al-Khwārizmī was a teacher in Baghdad who, like Alcuin, wrote his book as a means to educate people as to the importance of algebraic thinking. Because al-Khwārizmī was the first to treat algebra as an autonomous discipline, he has been designated the "father of algebra."

The *Propositiones* puts basic mathematics on display for everyone to enjoy. Engaging with the problems is, however, not only recreational or educational, but also intellectually enlightening, making it obvious that mathematical curiosity belongs to everyone, not just mathematicians, as does the pleasure of doing mathematics. The pleasure is, of course, different from the kind that comes from listening to, say, a work of great music, such as Beethoven's Ninth Symphony. One does not have to know musical theory, or composition techniques, in order to recognize the beauty and feel the emotional force of the symphony. A math problem cannot be characterized in the same way; it can only be experienced as challenging and ingenious. But this produces a form of pleasure nonetheless, evoked by the intellectual "kick" that results from discovering the hidden pattern that a problem might conceal.

This book could be used as an instructional manual in formal school settings, for self-study, or as a manual of recreational mathematics, given the many puzzles that it contains. By taking aspects of daily life and turning them into mathematical "propositions," Alcuin's text shows us that math contains within it the "principles of all things," as Aristotle so famously put it in his *Nicomachean Ethics*, written around 350 BCE.

Cited Works and Further Reading

Aristotle (350 BCE). *Nicomachean Ethics*, translated by W. D. Ross. Oxford: Oxford University Press, 2009.

Ascher, Marcia (1991). *Ethnomathematics: A Multicultural View of Mathematical Ideas*. Pacific Grove: Brooks.

Burkholder, Peter (1993). Alcuin of York's *Propositiones ad acuendos juvenes*: Introduction, Commentary & Translation. *History of Science & Technology Bulletin*, Vol. 1, number 2.

Chace, Arnold B. (1979). *The Rhind Mathematical Papyrus: Free Translation and Commentary with Selected Photographs, Transcriptions, Transliterations and Literal Translations*. Reston, VA: National Council of Teachers of Mathematics.

Dales, Douglas (2012). *Alcuin: Theology and Thought*. Cambridge: Cambridge University Press.

Fibonacci, Leonardo (1202). *Liber Abaci*, trans. by L. E. Sigler. New York: Springer, 2002.

Garver, Valerie L. (2017). Alcuin of York. *Oxford Bibliographies*. DOI: 10.1093/OBO/9780195396584-0221

Hadley, John and Singmaster, David (1992). Problems to Sharpen the Young. *Mathematics Gazette* 76: 102–126.

O'Connor, J. J. and Robertson, E. F. (2012). *Propositiones ad acuendos juvenes*. https://mathshistory.st-andrews.ac.uk/HistTopics/Alcuin_book/.

Pillis, John de and Rose, Nicholas (1988). *Mathematical Maxims and Minims*. Raleigh NC: Rome Press.

Singmaster, David (1998). The History of Some of Alcuin's *Propositiones*. In: P. L. Butzer, H. Th. Jongen, and W. Oberschelp (eds.), *Charlemagne and His Heritage 1200 Years of Civilization and Science in Europe*, Vol. 2, 11–29. Brepols: Turnhout.

1
Arithmetic

Prologue

The word *arithmetic* comes from the Greek *arithmētikē*, meaning "the art of counting." Arithmetic was a core subject in the ancient and medieval schools. Typically, the textbooks used to teach it would present arithmetical principles and techniques via appropriate problems, perhaps to instigate curiosity and motivate students to learn arithmetic more effectively. Alcuin's *Propositiones* falls into this tradition. It was tailored to cultivate mathematical inquisitiveness, rather than simply train students to carry out arithmetical operations mechanically.

Beginning with Thales (c. 624–546 BCE), one of the "Seven Wise Men of Greece," and the mathematician and seer Pythagoras (c. 580–500 BCE), arithmetic came to be seen as a study of the properties of numbers, not just an art of counting (McHale 1993). The formal study of these properties started with Euclid (c. 300 BCE) whose book, titled *Elements*, is considered to be the first comprehensive treatise on mathematics as an autonomous field of study. Although geometry is a major focus of Euclid's book, it contains sophisticated explanations and proofs that relate to arithmetic, including his famous proof of the infinity of prime numbers. As Alcuin clearly knew, one cannot enter the realm of mathematics without first acquiring knowledge of arithmetic.

The Problems

One could easily classify many of Alcuin's *Propositiones* as problems-puzzles in arithmetic or arithmetical thinking, but there are seven in particular (numbers 1, 15, 46, 49, 50, 52, and 53) in which arithmetic is specifically highlighted. The numbering of the problems in Alcuin's text is retained here (and throughout the remainder of this book), but the language, phrasing, measuring units, and the number system used by Alcuin (the Roman system) have been updated, or modified, to make them easier to understand.

1. *Problem of the Snail*

A snail was invited by a swallow to a lunch, which was a league away. Now, the snail could only crawl 1 inch per day. At this rate, how long did it take for the snail to crawl to that lunch? Note that 1 league = 1500 paces, 1 pace = 5 feet, 1 foot = 12 inches, and 1 year = 365 days.

15. *Problem of the Furrows*

When a farmer goes plowing to make furrows, he turns, in total, three times at each end of his field. So, given this pattern, how many furrows does the farmer make?

[Note that a furrow is a long trench made in the ground by a plow for planting or irrigation.]

46. *Problem of the Talents*

While a certain man was walking, he came across a small bag on the street. He picked it up finding that it contained a number of coins adding up to 2 talents. A crowd of people noticed that he had found the bag and said to him: "Friend, give us a part of the money." But the man shook his head and said "no." The crowd then rushed at him, emptying the coins out of the bag, with each person grabbing 50 coins. The man was then left with 50 coins as well after the crowd left. How many people were there in the crowd?

[A talent is an ancient unit. Assume that 1 talent is equal to 5400 coins.]

49. *Problem of the Seven Carpenters*

Seven carpenters each made seven wheels in order to build a number of carts. How many carts did they build altogether?

[Needless to say, a cart has four wheels.]

50. *Problem of the Pints*

How many pints of wine are there in 100 measures of wine? A measure equals 48 pints. Also, how many cups are there in 100 measures? A pint equals six cups.

[Note that the problem requires two answers.]

52. *Problem of the Camel and Measures of Grain*

The head of a household ordered that 90 measures of grain be moved from one of his houses to a second one 30 leagues away. A camel can carry the 90 measures in three trips. On the way, however, the camel eats one measure of grain for each league that it travels. Given this fact, how many measures were left after the camel delivered the grain to the second house?

53. *Problem of the Monks*

The abbot of a monastery, which had 12 monks, asked his treasurer to give an equal share of the 204 eggs he had stored away to each of the monks— five priests, four deacons, and three readers. How many eggs were given to the priests, the deacons, and the readers?

Solutions

1. *Answer*: 246 years and 210 days

Solution

First, we compute how many inches there are in one league. Since there are 1500 paces in 1 league and each pace is equal to 5 feet, therefore, the distance in terms of feet is 1500 × 5 = 7500 feet. Since there are 12 inches in a foot, a league is thus equal to 7500 × 12 = 90,000 inches. That is the distance which the snail must travel in order to get to the lunch. Since the snail travels at 1 inch per day and there are 365 days in a normal year, it will travel 365 inches in a year. So, to cover 90,000 inches the snail will need 90,000 ÷ 365 = 246.575 years, or 246 years and 210 days, to do so.

15. *Answer*: seven furrows

Solution

Alcuin actually gives the answer of six, reasoning as follows: one turn = one furrow, and so, because the farmer makes three turns at one end of the field and three at the other end, there will be six furrows in total. But there is another possibility, whose solution can be broken down as follows:

- The farmer plows his first furrow at the initial end. He does not take a turn there, since it is his starting point.
- After plowing the furrow, he goes straight down to the other end of the field, plows a second furrow there, and then makes his first turn to go back to the initial end.
- At the initial end, he plows his third furrow, and then turns for a second time to go to the other end.
- At that end, he plows his fourth furrow, after which he turns for the third time to go back to the initial end.
- At the initial end, he plows his fifth furrow, and then makes his fourth turn to go back to the other end.

- At that end, the farmer plows his sixth furrow, after which he turns for the fifth time in order to go back to the initial end.
- At the initial end, he plows his seventh furrow, and turns for the sixth time, but does not go to the other end of the field because he has made the six turns required by the problem, which states that he turned three times at each end, or six in total.

The solution can be summarized with symbol notation as follows (F = furrow, T = turn, n = number of a furrow or turn). The counting principle involved is that there will be one furrow more than the number of turns, because at the start there are no turns, even though a furrow is plowed nonetheless. This can be formulated as $F_n = T_{n-1}$, which states that there will be one less turn than a furrow.

$F_1, T_0 \rightarrow$ The farmer plows his first furrow, F_1, at the initial end of the field. There is no turn, since it is the starting point, so T is T_0.

$F_2, T_1 \rightarrow$ The farmer then goes directly to the other end, where he plows his second furrow, F_2, after which he makes his first turn, T_1, on his way back to the initial end of the field.

$F_3, T_2 \rightarrow$ While there, he plows his third furrow, F_3, after which he makes his second turn, T_2, to go back to the other end.

$F_4, T_3 \rightarrow$ While at that end, he plows his fourth furrow, F_4, and makes his third turn, T_3, to go back to the initial end.

$F_5, T_4 \rightarrow$ Back at the initial end, he plows his fifth furrow, F_5, and then makes his fourth turn, T_4, to go back to the other end.

$F_6, T_5 \rightarrow$ There, the farmer plows his sixth furrow, F_6, after which he turns for the fifth time, T_5, to go back to the initial end.

$F_7, T_6 \rightarrow$ At the initial end, he plows his seventh furrow, F_7, turns for the sixth time, T_6, but does not go to the other end of the field because he has made six turns, as required by the problem, which states that he turned three times at each end, which is six turns in total.

The solution can be modeled with a diagram (Figure 1.1). Note that the three turns at each end are shown by the curves on the diagram:

46. *Answer*: 215 people

Solution

One talent equals 5400 gold coins. Therefore, 2 talents is 5400 × 2 = 10,800. This is the number of coins originally in the bag. After each person in the crowd took 50 gold coins for themselves there were 50 gold coins left in the bag for the finder. So, if we subtract this from the number of coins originally in the bag, we will get the total number of coins taken by the

$F_1 (T_0)$ F_3 T_2 F_5 T_4 F_7 T_6 [End]

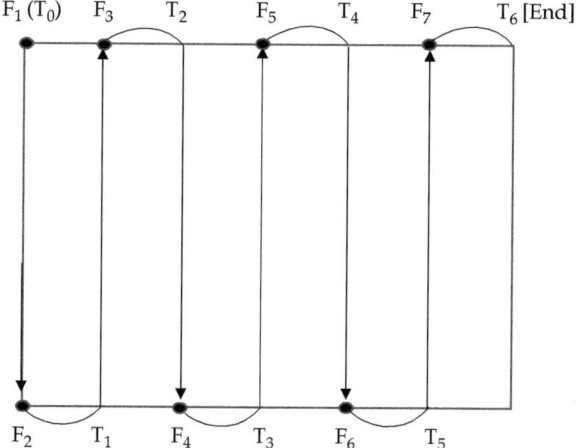

F_2 T_1 F_4 T_3 F_6 T_5

Figure 1.1 Model of the Solution to Problem 15

crowd: $10,800 - 50 = 10,750$. Since each person took 50 gold coins, the number of people can be determined by dividing this figure into 10,750, since 50 coins represent one person: $10,750 \div 50 = 215$.

49. *Answer*: 12 carts, with one wheel left over

Solution

There were seven carpenters who made seven wheels each, for a total of $7 \times 7 = 49$ wheels. Each cart requires four wheels, so the number of carts made is $49 \div 4 = 12$, plus one wheel left over.

50. *Answer*: 4800 pints and 28,800 cups

Solution

Since 1 measure = 48 pints, then 100 measures is equal to the following: $48 \times 100 = 4800$ pints. Since 1 pint = 6 cups, then 4800 pints is equal to the following: $6 \times 4800 = 28,800$ cups.

52. *Answer*: 20 measures

Solution

This is a true brainteaser. The solution can be broken down as follows:

- The camel is loaded with 30 measures of grain at the first house, leaving 60 of the 90 measures to be transported behind. This is so because the problem states that the camel took three trips, meaning that the maximum number of measures it was able to carry per trip was one-third of the 90 measures, which is 30 measures
- For the trip to be successful—that is, for any of the grain to reach the second house—the camel will need to make a stop at a location, say,

20 leagues from the start. Otherwise, if it does not make a stop, it will have eaten all of the 30 measures of grain by the time it reaches the second house, because it eats one measure per league, and there are 30 leagues to the second house.

- On its first trip to the location, the camel will have eaten 20 measures of the 30 on its back, leaving 10 measures at the location.
- The camel then goes back to the start and is reloaded with 30 more measures of the remaining 60 measures for its second trip, leaving the last 30 measures behind.
- The camel goes to the same location 20 leagues away. Again, it will have eaten 20 measures of the 30 on its back along the way, leaving another 10 measures at that location. So far, a total of 20 measures have been left at that location.
- The camel goes back to the start and is loaded up with the remaining 30 measures for its third trip.
- It goes to the location, again having eaten 20 measures along the way, leaving another 10 measures there—for a total of 30 measures. Those 30 measures are then loaded onto the camel to take to the second house.
- Now, since the second house is 30 leagues away from the first house, while the location is 20 leagues away from the first house, then the distance from the location to the second house is 10 leagues.
- Loaded with the 30 measures, the camel then makes its way to the second house 10 leagues away, eating 10 of the measures along the way, leaving 20 measures at the second house.

53. *Answer*: The priests receive 85 eggs, the deacons 68, and the readers 51

Solution

Each of the 12 monks will receive 204 ÷ 12 = 17 eggs, which represents an equal division of the eggs. Therefore:

- The five priests will receive a total of 5 × 17 = 85 eggs.
- The four deacons will receive a total of 4 × 17 = 68 eggs.
- The three readers will receive 3 × 17 = 51 eggs.
- Total: 85 + 68 + 51 = 204 eggs.

Annotations

The ancient Greek mathematicians were among the first to formally differentiate between counting and arithmetic as a discipline for establishing principles of counting and the properties of the numbers used in counting.

Among the seven problems above, there are several that stand out, bearing implications for the arithmetic that came subsequent to Alcuin. As an aside, it is hard to imagine today how Alcuin and his readers would have been able to regularly solve such problems with the cumbersome Roman numeral system, which was in use at the time.

Box 1.1 Roman Numerals

The Roman numeral system was based on seven alphabet letters having specific numerical values:

$$I = \text{one}$$
$$V = \text{five}$$
$$X = \text{ten}$$
$$L = \text{fifty}$$
$$C = \text{one hundred}$$
$$D = \text{five hundred}$$
$$M = \text{one thousand}$$

To grasp how unwieldy the system was, consider the number "two thousand two-hundred and fifty-three." In Roman numerals, it was represented as "MMCCLIII." Now, compare this with the numeral we use commonly today: "2,253 = MMCCLIII." The position of each digit in it indicates its value as a power of 10, which is why the system is called "decimal" (from Latin *decem* "ten"). Now, imagine trying to carry out a simple arithmetical task, such as adding 2253 + 1337 = 3590, with Roman numerals:

$$\text{MMCCLIII} + \text{MCCCXXXVII} = \text{MMMDXC}$$

Using Roman numerals is further complicated by the fact that a letter sign representing a smaller value, placed before a sign standing for a larger numerical value, is subtracted from the larger one. To wit: the numeral for "ninety" is represented by XC ("one hundred minus ten").

The Snail Problem

In Alcuin's era, reference to snails in problems was common, as was reference to serpents, both of which were animal metaphors. One of the most famous of these traveled from ancient India to Europe at around the same time as Alcuin's problem. It is paraphrased below:

> A snail is at the bottom of a 30-foot well. Each day it crawls up 3 feet and slips back 2 feet. The snail has the ability to stick to the walls of the well and, thus, does not slide down to the bottom at the end of a day when it stops to rest. At that rate, when will the snail be able to reach the top of the well?

Here is a breakdown of the snail's journey up the well:

> Day 1: Goes up to the 3-foot mark and slides down to the 1-foot mark.
> Day 2: Starts at the 1-foot mark, goes up to the 4-foot mark and slides down to the 2-foot mark.
> Day 3: Starts at the 2-foot mark, goes up to the 5-foot mark and slides down to the 3-foot mark.
> . . .
> Day 26: Starts at the 25-foot mark, goes up to the 28-foot mark and slides down to the 26-foot mark.
> Day 27: Starts at the 26-foot mark, goes up to the 29-foot mark and slides down to the 27-foot mark.

Consider the start of day 28. The snail finds itself at the 27-feet mark from the bottom. This means that the snail has 3 feet to go to the top on that day. It goes up the 3 feet, reaches the top, and crawls out. Versions of the same puzzle, with different details, such as a different animal at the bottom of the well and distance to the top, include one by Leonardo Fibonacci in his *Liber Abaci* of 1202 and a version found in an arithmetic textbook written by Christoff Rudolf, published in Nuremberg in the mid-1500s.

The Number Line

While the *number line* concept was introduced after the medieval era, its intellectual seeds can be seen in problems such as the snail puzzle, which involve moving along a linear path in terms of consistent numerical values. The movement "up" on the path traveled by the snail, for instance, represents a positive integer movement (which would be to the right of zero on the number line) and "down" represents a negative integer movement (to the left of zero on the number line). The first 10 integers on the number line, positive and negative, are shown in Figure 1.2. These are placed at equal distances—positive numbers to the right of zero and negative numbers to the left of zero:

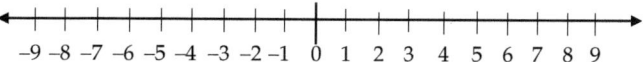

Figure 1.2 The Number Line

Zero is neither negative nor positive. It is a "boundary point" between the positive and negative numbers. All the real numbers (rational and irrational) can be placed in an exact way on the number line—hence its name as the "real number line" (Figure 1.3).

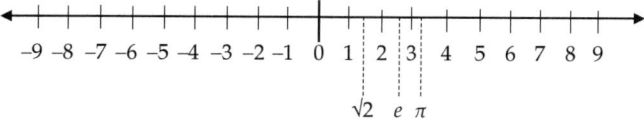

Figure 1.3 The Real Number Line

Box 1.2 Rational and Irrational Numbers

Rational numbers are those that can be represented with the ratio p/q, with p and q standing for integers, and $q \neq 0$. It stands for both the integers when $q = 1$: {2/1 = 2, 3/1 = 3, 4/1 = 4, etc.}; and the fractions: {1/2, 2/3, 3/4, 4/5, . . . }. By making p negative we also get all the negative rational numbers: $-p/q$ = {-1, -2, -1/2, -2/3, -3/4, -4/5, . . . }. *Irrational* numbers are those that cannot be expressed as p/q, such as the number $\sqrt{2}$.

An important irrational number is π *(pi)*, defined as the ratio of a circle's circumference (C) to its diameter (D): $\pi = C/D$, no matter how large or small the circle might be. The ancient Hebrews took the ratio to be 3. The Babylonians also thought it was 3; while the Egyptians estimated it to be 3.1604. Its actual value is: $\pi = 3.141592653589793$ . . . . The digits after the decimal go on forever without repeating. The number "e" is another important irrational number: e = 2.71828182845 . . . . It is called *Euler's number*, after Swiss mathematician Leonhard Euler (1707–1783), who referred to it in a 1731 letter and then again in his 1736 book, *Mechanica*.

The first conceptualization of the number line is found in John Wallis's (1616–1703) *Treatise of Algebra* (1685), in which he describes addition and subtraction in terms of a person moving forward and backward.

The Mystical Number Seven

Alcuin's problem of the seven carpenters (number 49), as straightforward as it seems, has a likely numerological aspect to it, of which Alcuin would certainly have been aware—that is to say, the number 7 was perceived to have mystical qualities, constituting a numerological tradition that went back to antiquity. Although Alcuin does not allude to this symbolism in any specific way, there is little doubt that his choice of the number 7 for his problem would have reverberated with this meaning among his medieval readers. The same number is found in a famous problem in the *Rhind Papyrus*—Problem 79.

Box 1.3 The Rhind Papyrus

This is one of the oldest math texts, dating to before 1650 BCE, and likely used as a textbook in Egyptian schools. It is called either the *Ahmes Papyrus*, after the Egyptian scribe, Ahmes, who copied it in that year, or the *Rhind Papyrus*, after the Scottish lawyer and antiquarian, A. Henry Rhind, who purchased it in 1858 while vacationing in Egypt. In addition to over 80 challenging problems (most of which are *de facto* puzzles), the *Papyrus* contains tables for the calculation of areas, methods for converting fractions, an analysis of elementary sequences, and information about measurement.

The problem presents an inventory without a question, implying, in all likelihood, that the readers had to figure out what hidden pattern was involved without any guidance that a question might provide:

Houses	7
Cats	49
Mice	343
Sheaves of wheat	2401
Hekats of grain	16, 807
Estate	19, 607

The problem seems intractable at first. However, by looking closely at the numbers representing the items of the estate (houses, cats, etc.), an arithmetical insight crystallizes—the first five numbers are successive powers of 7: $7 = 7^1$, $49 = 7^2$, $343 = 7^3$, $2401 = 7^4$, and $16,807 = 7^5$. But this pattern does not apply to the last figure, 19,607. Clearly, we need another hunch. Since the first five numbers represent items that are found in an estate, then the last number, 19,607, is probably the sum of these numbers, especially given the fact that it is placed at the end of the list, like the sum in any addition problem. And this is indeed the final touch to the solution: 7 + 49 + 343 + 2401 + 16,807 = 19,607.

The mystical fascination with the number 7 may be the reason why the same problem blueprint surfaces elsewhere and in different eras, as for instance, in Fibonacci's *Liber Abaci* of 1202. Fibonacci used different items, as is to be expected, added another power of 7, namely 7^6, and directly asked for the sum of all the items, unlike the Egyptian one, which had to be inferred:

> Seven old women are on the road to Rome. Each woman has seven mules, each mule carries seven sacks, each sack contains seven loaves, to slice each loaf there are seven knives, and for each knife there are seven sheaths to hold it. How many are there altogether, women, mules, sacks, loaves, knives, sheaths?

The answer is: $7^1 + 7^2 + 7^3 + 7^4 + 7^5 + 7^6 = 137,256$. Fibonacci could not possibly have known about the Egyptian counterpart, because the existence of the *Papyrus* was not known during his times, nor had hieroglyphic writing been deciphered yet. But the similarity between the two is unmistakable. Since Fibonacci spent his childhood in North Africa where his father was a customs officer and was educated there by the Moors, he may, however, have heard oral versions of the puzzle in his childhood.

The number 7 has had a strange appeal across time and cultures. In the Bible, the world was created in seven days and seven nights; there are seven gods of good fortune in Japanese lore; seven chieftains in Greek mythology undertook an ill-fated expedition against the city of Thebes; there are seven deadly sins, according to medieval theologians; there are seven cosmic truths according to the Plains Peoples of North America; and the list could go on and on. In most ancient mathematical traditions, the distinction between *numeration* (numbers in themselves) and *numerology* (the philosophical and metaphysical meanings that they are believed to bear) is almost never maintained.

The Jeep Problem

Alcuin's problem of the camel (number 52) is the first known appearance of what has come to be known as an optimization problem (Fine 1947, Gale 1970). A well-known version is called the "jeep problem," adapted and simplified here for illustration purposes:

A jeep has to carry a person across the desert to a hospital. There is no gas station in the desert and the jeep has just enough fuel to get it half way across the desert. The jeep consumes 1/3 of its fuel per 1/6 unit distance. There are three other jeeps who accompany the jeep with the patient, so that they can transfer their fuel into it. All the four jeeps consume fuel at the same rate. How can the person be driven across the desert to the hospital?

The solution can be broken down as follows:

- At 1/6 unit distance from the start, each jeep will have consumed 1/3 of its fuel, leaving 2/3 in each tank.
- One of the jeeps transfers its remaining 2/3 fuel equally to two of the others: 1/3 to each one. These are now again full, while a third jeep remains only 2/3 full. The empty jeep stays behind and will be rescued later. The other three continue on their way through the desert.
- At a subsequent 1/6 distance, the two full jeeps become again 2/3 full, consuming 1/3 fuel each. The one that was 2/3 full also consumes 1/3 fuel, leaving 1/3 in its tank. This is used to fill up the jeep with the patient in it (2/3 + 1/3 = 1). The other jeep remains 2/3 full. The two jeeps continue on their journey. The empty one will be rescued later.
- At a subsequent 1/6 distance, the middle of the desert is reached, since together the three distances from the start add up to one half: 1/6 + 1/6 + 1/6 = 3/6 = 1/2.
- At that point, the jeep with the patient loses 1/3 fuel, as per its consumption rate, leaving 2/3 in its tank. The other jeep, which had 2/3 fuel in the tank, consumes 1/3 fuel as well, leaving 1/3 in its tank. This is transferred to the jeep with the patient, filling it up (2/3 + 1/3 = 1). The empty jeep will be rescued later.
- The jeep with the patient is now full. Finding itself halfway to the end of the desert, it can now make the rest of the journey to the hospital with the fuel it has, since 1/2 the distance is equal to 1/6 + 1/6 + 1/6, which are the intervals at which the jeep loses 1/3 the fuel it has. It has a full tank, so it will use it all up, 1/3 + 1/3 + 1/3 = 1, getting it to the hospital.

Tartaglia's Problem

Alcuin's choice of a camel for his problem is a culturally interesting one. Not only were these animals used as a means of travel and transportation in his era, especially across long distances, but they also had monetary value in an inheritance. An ingenious arithmetical puzzle, traced (but unverified) to Italian mathematician Niccolò Tartaglia (1499–1557), involves this latter meaning. It is paraphrased as follows:

> A man dies, leaving 17 camels to be divided among his heirs, in the proportions 1/2, 1/3, 1/9. How can this be done?

Dividing up the camels in the manner decreed by the father would entail having to split up a camel. This would, of course, kill it. So, Tartaglia suggested "borrowing an extra camel," for the sake of argument, not to mention humane purposes. With 18 camels, he arrived at a practical solution: one heir was given 1/2 (of 18), or 9, another 1/3 (of 18), or 6, and the last one 1/9 (of 18), or 2. The 9 + 6 + 2 camels given out in this way add up to the original 17: 9 + 6 + 2 = 17. The extra camel could then be returned to its owner.

Whether or not this is a legal solution is beside the mathematical point. The puzzle bears a general implication:

- If there are n camels, and n is an odd number, 3 heirs, and ratios $1/a$, $1/b$, $1/c$, with $\{a, b, c\}$ referring to the denominators in the ratios, then by adding an extra camel, $(n + 1)$, the ratios divide evenly, producing integer solutions:

$$1/a \times (n + 1) + 1/b \times (n + 1) + 1/c \times (n + 1)$$

- The chart below shows the solutions for $n = 7, 11, 17, 19, 23, 41$.

Table 1.1 Integer Solutions for $n = 7, 11, 17, 19, 23, 41$.

Camels n	Denominators $\{a, b, c\}$	Proportions $\{1/a, 1/b, 1/c\}$	Proportions with extra camel $(n + 1)$
7	$a = 2, b = 4, c = 8$	1/2, 1/4, 1/8	$1/2 \times 8 = 4, 1/4 \times 8 = 2, 1/8 \times 8 = 1$ $\rightarrow 4 + 2 + 1 = 7$
11	$a = 2, b = 4, c = 6$	1/2, 1/4, 1/6	$1/2 \times 12 = 6, 1/4 \times 12 = 3, 1/6 \times 12 = 2$ $\rightarrow 6 + 3 + 2 = 11$
11	$a = 2, b = 3, c = 12$	1/2, 1/3, 1/12	$1/2 \times 12 = 6, 1/3 \times 12 = 4, 1/12 \times 12 = 1$ $\rightarrow 6 + 4 + 1 = 11$

Continued

Table 1.1 *Continued*

Camels n	Denominators {a, b, c}	Proportions {1/a, 1/b, 1/c}	Proportions with extra camel (n + 1)
17	$a = 2, b = 3, c = 9$	1/2, 1/3, 1/9	$1/2 \times 18 = 9, 1/3 \times 18 = 6, 1/9 \times 18 = 2$ → 9 + 6 + 2 = 17
19	$a = 2, b = 4, c = 5$	1/2, 1/4, 1/5	$1/2 \times 20 = 10, 1/4 \times 20 = 5, 1/5 \times 20 = 4$ → 10 + 5 + 4 = 19
23	$a = 2, b = 3, c = 8$	1/2, 1/3, 1/8	$1/2 \times 24 = 12, 1/3 \times 24 = 8, 1/8 \times 24 = 3$ → 12 + 8 + 3 = 23
41	$a = 2, b = 3, c = 7$	1/2, 1/3, 1/7	$1/2 \times 42 = 21, 1/3 \times 42 = 14, 1/7 \times 42 = 6$ → 21 + 14 + 6 = 41

Note that when $n = 17$, we get Tartaglia's original puzzle. The question becomes: Is this really a solution to the division of inheritance? Or is it just a clever mathematical maneuver?

Epilogue

Alcuin's problems in arithmetic show, above all else, that there is more to it than just adding, subtracting, multiplying, and dividing. Such problems involve thinking about quantitative phenomena in an abstract way—a form of thought which often leads to insights that may not have been evident before engaging with them. Alcuin's problems clearly bring out that arithmetic is not a boring mechanical activity, but provides a unique way of looking at things, involving a search for some hidden numerical pattern.

In other words, arithmetic is everywhere one looks, from figuring out how to divide money, determining ways to plow a field, figuring out how to transport something under limiting conditions, and so on. As the poet Carl Sandberg once cleverly put it: "Arithmetic is numbers you squeeze from your head to your hand to your pencil to your paper till you get the answer" (citation from: https://allpoetry.com/Arithmetic).

Explorations

In this, and all subsequent chapters, a set of 10 problems is provided. These are either based imitatively on Alcuin's problems discussed in the chapter or else are meant to be illustrative of the ideas discussed in the Annotations. Some of the puzzles are classic ones in recreational mathematics; so slightly different versions of these may be found in other collections. Answers and solutions are given at the back.

1. *A Cow and Goat Problem*

Yesterday, a farmer bought a cow and a goat at a market square. Together they cost him 55 gold coins. The cow cost 50 coins more than the goat. How much did each one cost? For the purposes of this problem assume that gold coins come in both unit value and half unit value.

[This is a classic puzzle in recreational mathematics that has been reframed here as a problem in the style of Alcuin.]

2. *The Achilles and the Tortoise Paradox*

It is reported that the legendary hero, Achilles, decided to race against a tortoise. To make the race fairer, he allowed the tortoise to start at half the distance away from the finish line. In this way, however, Achilles will never surpass the tortoise. Why?

Box 1.4 Zeno's Paradoxes

This is one of the famous paradoxes attributed to the philosopher Zeno of Elea (c. 490–430 BCE), which he apparently devised in support of the doctrine of his teacher, Parmenides, that motion is an illusion. The paradox is recounted in Aristotle's *Physics* (VI: chapter 9). It is included here as an exploration because it involves segmenting a linear path in a way that leads to an apparent paradox. It also encapsulates what Alcuin was trying to do with his own book overall—to illustrate that mathematical thinking is more than simple calculation. In 1895, Lewis Carroll wrote his own version of the paradox, implying that Zeno's conundrum was not about time and space, but rather about the nature of reasoning itself—a theme adopted by Douglas Hofstadter in his 1979 book, *Gödel, Escher, Bach: An Eternal Golden Braid*. Hofstadter also sees the paradox as a challenge to logic itself—a challenge that, he claims, was taken up in 1931 by Kurt Gödel with his famous incompleteness theorem (discussed in the next chapter).

---

3. *The Von Neumann Problem*

A boy and a girl were out riding their bikes yesterday, coming at each other from opposite directions. When they were exactly 20 miles apart, they began racing toward each other. The instant they started doing so, a fly on the handle of the girl's bike started flying toward the boy. As soon as it reached the handle on his bike, it turned around and started back toward the girl. The fly flew back and forth in this way, from handle bar to handle bar, until the two bicycles met. Each bike moved at a constant speed of 10 miles an hour, and the swifter fly flew at a constant speed of 15 miles an hour. How much total distance did the fly cover?

Box 1.5 John von Neumann (1903–1957)

This puzzle is part of a legendary anecdote told about John von Neumann, the Hungarian-born professor of mathematics at Princeton University, whose ideas were instrumental to the development of the modern computer. According to one version, the puzzle was posed to him at some gathering by a student. Von Neumann thought about it for a moment and then stated the correct answer. The student who posed the puzzle was amazed at the instant response, remarking that highly skilled mathematicians tended to overlook the simple way in which the puzzle was solved, trying instead to solve it by a lengthier process of summing a series. Von Neumann is said to have looked quizzically at the fellow, retorting matter-of-factly, "Well, that's precisely how I solved it." This puzzle is included here because like many of Alcuin's problems it illustrates that mathematics often involves ingenious spatial reasoning.

---

4. *The Warehouse Fire Puzzle*

During a warehouse fire, a firefighter stood on the middle rung of a ladder, pumping water into the burning warehouse. A minute later, she stepped up three rungs and continued directing water at the building from her new position. A few minutes after that, she stepped down five rungs, and from her new position continued to pump water into the building. Half an hour later, she climbed up seven rungs and pumped water from her new position until the fire was extinguished. She then climbed the remaining seven rungs up to the roof of the warehouse. How many rungs were on the ladder?

[This is a classic puzzle in recreational mathematics, used here to emphasize the utility of the concept of the number line, as discussed in the Annotations.]

5. *The Train Passing Problem*

A train leaves New York for Washington every hour on the hour. Similarly, a train leaves Washington for New York, but it does so every half-hour. The trip takes 5 hours each way. If we were on the train from New York bound for Washington, how many of the trains coming from Washington and going toward New York would we pass?

[This is yet another classic problem that is used here to bring out, again, the practical importance of the concept of the number line (Annotations).]

6. *A Take on Alcuin's Problem 46*

A person comes across a bag on the street. She picks it up and sees that it contains gold coins. She counts them and, being a mathematician, utters the following to herself. "Well, I see that the coins add up to a three-digit number, and that all three digits in the number are the same. If I were to divide the number in half, it would produce another three-digit number whose digits are identical, and when added together, equal 3." How many coins were in the bag?

Box 1.6 Numbers and Numerals

A *number* is a quantity or value. A *numeral* is a sign that represents a number. For example, the number (quantity) "seven" is represented with the numeral "7" in the decimal system, with "VII" in the Roman system, or with "111" in the binary system. The numeral system we use commonly today is the decimal one, based on 10 digits (meaning "fingers" in Latin). It is also called, logically, a base-ten system.

---

7. *A Mountain Climbing Problem*

Danielle is a mountain climber. She can hike uphill at an average rate of 2 miles per hour, and downhill at an average rate of 6 miles per hour. Going uphill and downhill, without stopping at the top, what will her average speed for a whole trip be?

[This type of problem is found commonly in both school textbooks and puzzle collections. It is used here to emphasize the importance of thinking clearly in arithmetic, since the solution could be easily overlooked—which was one of Alcuin's own pedagogical objectives.]

8. *A Coin Puzzle*

My friend Jim has 20 American coins in his pocket, consisting of dimes and nickels. Altogether the coins add up to $1.35. Note that a dime is equal to

10 cents, or $0.10 of a dollar, and a nickel to 5 cents, or $0.05 of a dollar. How many of each does he have?

[This type of problem is also found commonly in school texts and puzzle anthologies. It exemplifies how even a seemingly simple counting activity might still involve ingenious arithmetical reasoning.]

9. *A Work Rate Problem*

It takes Beatrice twice as long as it takes Amber to do a certain piece of work. Working together, they can do the work in six days. How long would it take each one to do it alone?

[This type of arithmetical puzzle traces its origins to the *Greek Anthology* (see Introduction). To this day, it is found in elementary arithmetic texts in schools across the world.]

10. *Sam Loyd's Archery Puzzle*

The other day, at an archery competition, a young lady, who carried off the first prize, scored exactly 100 points. The scores on the target itself are given as: 16, 17, 23, 24, 39, and 40. Can you figure out how many arrows she must have used to accomplish the feat?

Box 1.7 Sam Loyd (1841–1911)

This puzzle was devised by American puzzle-maker Sam Loyd (1959). Loyd was born in Philadelphia but grew up in New York City. He was one of the first individuals to earn a comfortable living from puzzle-making alone. As Matthew J. Costello describes him, in his book, *The Greatest Puzzles of All Time* (1988), Loyd was "puzzledom's greatest celebrity, a combination of huckster, popularizer, genius, and fast-talking snake oil salesman." Loyd's interest in puzzles apparently started in 1860, when he became problem editor of the magazine *Chess Monthly*. By 1878, Loyd started creating chess puzzles. So popular did these become among his readers that he became convinced he could make a decent living working as a puzzle-maker, even though he was trained as an engineer. It is included here because, like many of Alcuin's problems, it requires a lot of arithmetical ingenuity to solve.

Cited Works and Further Reading

Ball, W. W. Rouse and Coxeter, H. S. M. (1987). *Mathematical Recreations and Essays*, 13th edition. New York: Dover.

Burkholder, Peter (1993). Alcuin of York's *Propositiones ad acuendos juvenes*: Introduction, Commentary & Translation. *History of Science & Technology Bulletin*, Vol. 1, number 2.

Cajori, Florian (1909). *A History of Mathematics*. London: Macmillan.

Carroll, Lewis (1895). What the Tortoise Said to Achilles. *Mind* 4: 278–280.

Chace, Arnold Buffum (1979). *The Rhind Mathematical Papyrus: Free Translation and Commentary with Selected Photographs, Transcriptions, Transliterations and Literal Translations*. Reston: National Council of Teachers of Mathematics.

Euler, Leonhard (1736). *Mechanica*. Petropoli: Typographia Academiae Scientiarum.

Fibonacci, Leonardo (1202). *Liber Abaci*, trans. by L. E. Sigler. New York: Springer, 2002.

Fine, N. J. (1947). The Jeep Problem. *American Mathematical Monthly* 54: 24–31.

Gale, David (1970). The Jeep Once More or Jeeper by the Dozen. *American Mathematical Monthly* 77: 493–501.

Gillings, Richard J. (1972). *Mathematics in the Time of the Pharaohs*. Cambridge, Mass.: MIT Press.

Hofstadter, Douglas (1979). *Gödel, Escher, Bach: An Eternal Golden Braid*. New York: Basic.

Loyd, Sam (1959–1960). *Mathematical Puzzles of Sam Loyd*, 2 volumes, compiled by Martin Gardner. New York: Dover.

Oberschelp, Walter (1995). Alcuin's Camel and the Jeep Problem. In: *Charlemagne and His Heritage*, 411–422. Turnhout: Brepols.

Maor, Eli (1998). *Trigonometric Delights*. Princeton: Princeton University Press.

McHale, Desmond (1993). *Comic Sections: The Book of Mathematical Jokes, Humor, Wit, and Wisdom*. Dublin: Boole Press.

Peet, Thomas E. (1923). *The Rhind Papyrus*. Liverpool: University of Liverpool Press.

Petkovic, Miodrag S. (2009). *Famous Puzzles of Great Mathematicians*. Providence, RI: American Mathematical Society.

Singmaster, David (2021). *Adventures in Recreational Mathematics*. Singapore: World Scientific.

Wallis, John (1655). *Arithmetica Infinitorum*. Leon: Lichfield.

2
Countability

Prologue

Some of the problems in Alcuin's *Propositiones* are ingenious exercises in the use of different methods of countability. One of these, problem number 41, regards exponential growth, and thus constitutes a type of problem that eventually led to the need to develop formal methods of exponentiation. As a prelude to the discussion of exponents, consider multiplying the number 3 fifteen times:

$$3 \times 3 \times 3 \times 3 \times 3 \times 3 \times 3 \times 3 \times 3 \times 3 \times 3 \times 3 \times 3 \times 3 \times 3 = 14,348,907$$

This layout is discernibly unwieldy. Even just looking at it boggles the mind for most of us. To make such problems easier to read, in the early Renaissance mathematicians came up with the concept of *exponent*, written as a superscript number, "3^{15}." In this notation, the "3" is called the *base*, and the superscript "15" the *exponent* or *power*. The exponent tells us the number of times that the base is to be multiplied. So, "3^{15}" is shorthand for the multiplication above:

$$3^{15} = (3 \times 3 \times 3 \times 3 \times 3 \times 3 \times 3 \times 3 \times 3 \times 3 \times 3 \times 3 \times 3 \times 3 \times 3) = 14,348,907$$

In general, "n^m" indicates that any number "n" is to be multiplied "m" times:

$$n^m = \underbrace{n \times n \times n \times n \ldots \times n}_{m \text{ factors}}$$

Although the idea of exponentiation as a technique for representing very large numbers goes back to Archimedes' work, *The Sand Reckoner* (c. 216 BCE), the exponential notation we use today was introduced by French mathematician Nicolas Chuquet (1445–1500), and refined later by German mathematicians Henricus Grammateus (1495–1526) and Michael Stifel (1487–1567). It was Stifel who coined the term *exponent*. It is fascinating to note that, right after such notation was introduced, it became the basis for stimulating new ideas, leading to different branches of mathematics (discussed in this chapter).

The Problems

Problems 6, 35, 41, and 43 in Alcuin's text involve notions of countability, hypothesis testing, exponential growth, impossibility, and decidability—all central ones in mathematical theory. So, while these may seem to be problems in simple arithmetic on the surface, they are actually gems in their own right, prefiguring later abstract mathematics.

6. *Problem of the Farmers and Pigs*

Two farmers had $100 between them, with which they bought groups of five pigs at $2 a group. The farmers intended to fatten the pigs so as to be able to sell them at a profit. But when they saw that the time was not right, and being unable to keep the pigs on their farms, they tried to make a profit by selling them quickly instead. However, they were unsuccessful because they could only sell the pigs for what they had paid (namely, five pigs for $2). So, they said to each other: "Let's divide the pigs into two (equal) groups, and sell them for as much as we had paid, after which we can actually make a profit." How many pigs were there at first, and how did the farmers divide and sell them for a profit?

35. *Problem of the Dying Man's Will*

A father died, leaving behind his pregnant wife and $960 in his estate. In his will, the father stipulated that if a son should be born to his wife, then the son should receive three-quarters of the $960. Moreover, if that happens, the mother should then get a quarter of the estate. However, if a girl were born instead, then the daughter should receive seven-twelfths of the estate, and in such case, the mother should receive five-twelfths. Now, as it turned out, the wife gave birth to twins, one boy and one girl. How much did the mother, son, and daughter each receive?

Box 2.1 Culture

There are several cultural peculiarities that this problem reflects, from a modern-day standpoint, which can be encapsulated by several questions. What about the other children left behind? Do they not get any inheritance? Why is there a discrepancy between the children when it comes to inheritance? All that can be said is that cultural factors crop up throughout the *Propositiones*, as they do in other math texts of the era. One should not, however, evaluate a text such as this one solely with modern eyes—all that can be said is that the times were different.

---

41. *Problem of the Sow and the Pigpen*

A farmer built a large square pigpen in which he placed a sow. The sow gave birth to seven piglets in the center of the pen. The offspring, along with the mother, then gave birth a little later to another seven piglets each in the first corner of the pigpen. After this, the sow and all the offspring in the pen each gave birth to seven more piglets in the second corner. The same happened in the third corner, and then in the fourth corner. Finally, the sow and all the offspring each gave birth to seven more piglets in the center of the pigpen. How many pigs, including the mother, were in the pigpen by this time?

[Note that Alcuin does not tell us how the sow became pregnant over and over, nor how the female piglets did as well. Perhaps in the medieval period matters of this kind were best left unsaid. So, one must disregard the implausible scenario depicted by the problem in terms of pig reproduction, and just read it for its mathematical content.]

43. *Problem of the Pigs*

A man had 300 pigs. He ordered that all of them be slaughtered in three days, but with an odd number to be killed on each day. He wished the same thing to be done with just 30 pigs. What odd number of pigs (of 300 or 30) were to be killed on each of the three days?

Solutions

6. *Answer:* 250 pigs, and see solution below for how the farmers made a profit

Solution

The breakdown is as follows. There are also other ways to solve the problem.

- The farmers paid $2 for one group of five pigs.
- Since the farmers paid $100 in total, we can deduce that there were 50 groups of pigs, each group costing $2: $2 × 50 = $100.
- Since there were five pigs per group, and 50 groups, the total number of pigs was: 5 × 50 = 250.
- We are told that the farmers sold the pigs for as much as they had paid, that is, $100 in total, and also made a profit. Here's how they did it.
- They divided the 250 pigs into two equal groups of 125 pigs each, as we are told. Let us call them A and B.

- They sold 120 of group A (not 125) at $1 per three pigs. How many such triplets were there? There were 120/3 = 40. At $1 each, this came to $1 × 40 = $40. This left them with five pigs from group A.
- They sold 120 of group B at $1 per two pigs. There were 120/2 = 60 such doublets. At $1 each, this came to 1 × 60 = $60. This left them with another five pigs from group B.
- All told, they still had 10 pigs to sell, after making back their $100 with the previous transactions: $40 + $60 = $100. They could then sell the 10 pigs left over, thus making a profit.

35. *Answer:* mother $320, son $360, daughter $280

Solution

The breakdown is as follows:

- If only a son were born, he would receive 3/4 of the inheritance, which is 3/4 × 960 = 720.
- The mother would then get 1/4 × 960 = 240.
- If only a daughter were born, she would receive 7/12 of the inheritance, which is 7/12 × 960 = 560.
- The mother would then get 5/12 of the estate, which is 5/12 × 960 = 400.
- Now, as it happened, twins of different gender were born.
- So, the twin son, being one of two children, gets half of what he would get as an only child: 1/2 × 720 = 360.
- The twin daughter, also being one of two children, gets half of what she would get as an only child: 1/2 × 560 = 280.
- The mother receives what is left over: 960 − (360 + 280) = 960 − 640 = 320.

41. *Answer:* 262,144 pigs

Solution

This problem is about powers of 8. It can be broken down as follows:

- After the sow had her first seven piglets, there were eight pigs in total in the pigpen (including the sow).
- This can be represented with the power of 1, $8^1 = 8$ pigs.
- Each of the seven piglets then had seven piglets of their own, as did the mother again, in the first corner.
- So, the eight pigs in the pen (with the sow) produced another eight pigs each, or 8 × 8 = 64 pigs, which is how much were then in the pen.
- This can be represented with the power of 2, $8^2 = 64$ pigs.
- Continuing to reason in this way:

Addition of pigs to the pen from the second corner:

8^3 pigs = 512 pigs

Addition of pigs to the pen from the third corner:

8^4 pigs = 4096 pigs

Addition of pigs to the pen from the fourth corner:

8^5 pigs = 32,786

Addition of pigs to the pen from the center:

8^6 pigs = 262,144, which is the total number of pigs in the pen.

43. *Answer:* no solution

Solution

The problem has no solution since the sum of three odd numbers can never be even. Alcuin knew that the problem was insoluble, since he remarked in the text that it was a good problem to give to children who had been naughty.

The problem asks that over three days, three odd numbers of pigs should be slaughtered, adding up to 30 or 300, which are even numbers. So, it is impossible because three odd numbers added together will never produce an even sum. Take a few odd random triplets as examples:

$$3 + 5 + 7 = 15 \ \text{(an odd number)}$$
$$11 + 17 + 23 = 51 \ \text{(an odd number)}$$
$$9 + 15 + 19 = 43 \ \text{(an odd number)}$$

Consider the case of consecutive odd numbers. If we let the first odd number be $(2n + 1)$, then the second odd number is $(2n + 3)$, and the third $(2n + 5)$. Such expressions will be discussed below. When added together these will add up to an odd number:

$$(2n + 1) + (2n + 3) + (2n + 5) = 6n + 9$$

As a concrete example, let us say that $n = 5$; then:

$$6n + 9 = 6\,(5) + 9 = 39 \ \text{(an odd number)}$$

Let us try another example, namely $n = 8$:

$$6n + 9 = 6\,(8) + 9 = 57 \ \text{(an odd number)}$$

No matter what value n has in $6n + 9$, the expression will always produce an odd number. The same type of analysis can be used in the case of non-consecutive odd numbers.

Annotations

Alcuin's problem number 6 is a thought experiment in countability: How is something countable in specific ways (according to particular constraints)? The problem is an original one in recreational mathematics (and arguably in mathematics overall), since no similar one has been found prior to Alcuin. The *counting* numbers, or *integers*, allow us to represent quantities that literally can be "counted" as whole things, rather than as fractions (parts of things) or as some other kind of object. Now, what happens when the counting goes on to infinity? This is a problem that traces its roots to antiquity, as will be discussed below.

The word *integer* comes from a Latin word meaning "whole." Historically, only positive integers were considered to be valid integers—hence, their designation as the "natural" counting numbers. This definition changed over time to include negative numbers. Zero is defined as a "neutral" integer because it is neither negative nor positive.

Cardinality

In his book, *Two New Sciences* (1638), Italian scientist Galileo included a truly remarkable observation: When the set of square integers is matched, one-by-one, to the complete set of positive integers, the result is that there seem to be as many square integers as there are integers overall. This appears to be a paradox since the square integers are themselves a subset of the entire set of integers (Figure 2.1).

1	2	3	4	5	6	7	8	9	10	11	12	...
↕	↕	↕	↕	↕	↕	↕	↕	↕	↕	↕	↕	
1	4	8	16	25	36	49	64	81	100	121	144	...

Figure 2.1 Galileo's Correspondence Observation

No matter how far one continues matching the numbers in this way, there will never be a gap between the top and the bottom sets. This suggests that the "number" of all positive integers and the "number" of those in one of its proper subsets, the square integers, is the same. If one stops to think about it, this demonstration was a truly astounding one, suggesting that when we deal with infinity, things truly change.

In the 1870s, this type of demonstration was expanded by German mathematician Georg Cantor (1845–1918), who showed, for example, that all the subsets of the integers can be put in a one-to-one correspondence with the complete set of integers. Below are the even numbers and odd numbers put in a one-to-one correspondence with all the integers (Figure 2.2).

1	2	3	4	5	6	7	8	9	10	11	12	...
$\updownarrow$	$\updownarrow$	$\updownarrow$	$\updownarrow$	$\updownarrow$	$\updownarrow$	$\updownarrow$	$\updownarrow$	$\updownarrow$	$\updownarrow$	$\updownarrow$	$\updownarrow$	
2	4	6	8	10	12	14	16	18	20	22	24	...
$\updownarrow$	$\updownarrow$	$\updownarrow$	$\updownarrow$	$\updownarrow$	$\updownarrow$	$\updownarrow$	$\updownarrow$	$\updownarrow$	$\updownarrow$	$\updownarrow$	$\updownarrow$	
3	5	7	9	11	13	15	17	19	21	23	25	...

Figure 2.2 One of Cantor's Correspondence Observations

To elaborate his new vision of countability, Cantor introduced the notion of *cardinality*, adopting the term *cardinal* numbers that emerged in the sixteenth century to refer to all the positive counting numbers. With his brilliant demonstrations, Cantor showed that any subset of numbers that can be put in an infinite one-to-one correspondence with the set of integers will have the same cardinality, which he represented with the Hebrew letter $\aleph_0$ (aleph null). Cantor also proved, again amazingly, that the entire set of rational numbers has the cardinality $\aleph_0$. To demonstrate this, he devised a proof that has come to be known as his "diagonal proof." It is paraphrased as follows.

First, Cantor laid out all possible positive rational numbers, p/q, in a layout, now called "Cantor's sieve," in terms of a specific pattern:

- In each row, the successive denominators (q) of the numbers are the integers in order $\{1, 2, 3, 4, 5, 6, \ldots\}$.
- The numerator (p) of the numbers in the first row is 1, of those in the second row 2, of those in the third row 3, and so on.
- If every number in a row that is a multiple of another one in the sieve is skipped (underlined below), then every rational number, p/q, appears once and only once in the sieve (Figure 2.3).

$p = 1, q = \{1, 2, 3, \ldots\} \rightarrow$	1/1	1/2	1/3	1/4	1/5	1/6	1/7	1/8	...
$p = 2, q = \{1, 2, 3, \ldots\} \rightarrow$	2/1	<u>2/2</u>	2/3	<u>2/4</u>	2/5	<u>2/6</u>	2/7	<u>2/8</u>	...
$p = 3, q = \{1, 2, 3, \ldots\} \rightarrow$	3/1	3/2	<u>3/3</u>	3/4	3/5	<u>3/6</u>	3/7	3/8	...
$p = 4, q = \{1, 2, 3, \ldots\} \rightarrow$	4/1	<u>4/2</u>	4/3	<u>4/4</u>	4/5	<u>4/6</u>	4/7	<u>4/8</u>	...
$p = 5, q = \{1, 2, 3, \ldots\} \rightarrow$	5/1	5/2	5/3	5/4	<u>5/5</u>	5/6	5/7	5/8	...
$p = 6, q = \{1, 2, 3, \ldots\} \rightarrow$	6/1	<u>6/2</u>	<u>6/3</u>	<u>6/4</u>	6/5	<u>6/6</u>	6/7	<u>6/8</u>	...
$p = 7, q = \{1, 2, 3, \ldots\} \rightarrow$	7/1	7/2	7/3	7/4	7/5	7/6	<u>7/7</u>	7/8	...
$p = 8, q = \{1, 2, 3, \ldots\} \rightarrow$	8/1	<u>8/2</u>	8/3	<u>8/4</u>	8/5	<u>8/6</u>	8/7	<u>8/8</u>	...
and so on	...	...	...	...	...	...	...	...	...

Figure 2.3 Cantor's Sieve

Now, Cantor did something truly unique. He connected the numbers in the sieve with arrows that move diagonally, as shown below, which allowed him to put the rational numbers in the sieve in a one-to-one correspondence with all the integers (Figure 2.4).

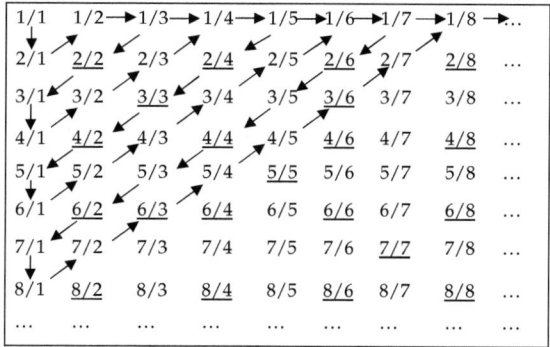

Figure 2.4 Cantor's Diagonal Proof

Here is the proof:

- The 1/1 at the top left-hand corner of the sieve is put in correspondence to the integer 1.
- Following the arrow down, the 2/1 below is put in correspondence to the integer 2.
- Following the arrow diagonally, the 1/2 at the top is put in correspondence to the integer 3.
- Following the arrow, the next number is 1/3, which is put in correspondence to the integer 4.
- We skip the 2/2 below, since it equals 1, as does the previous 1/1.
- The next diagonal number is 3/1, and this is put in correspondence to the integer 5
- And so on, throughout the sieve.
- The path indicated by the arrows thus allows us to set up a one-to-one correspondence between the integers and the rational numbers (Figure 2.5).

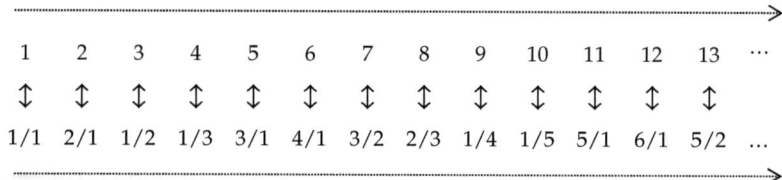

Figure 2.5 Correspondence between the Integers and the Rationals

This shows that the rational numbers have the cardinality of $\aleph_0$. Cantor's proofs and ideas behind them became the basis for the modern day notion of sets, defined as collections of numbers sharing a property, thus catapulting mathematics considerably into its modern era.

Hypothetical Thinking

Solving Alcuin's inheritance problem (number 35) involves a form of hypothetical thinking that has the following general logical structure: "if *a*, then *b*; if *c*, then *d*." If we replace *a* with the statement "if a son should be born," and *b* with "then the son should receive three quarters of the 960 dollars," *c* with "if a girl were born instead," and *d* with "then the daughter should receive seven twelfths of the estate," we can see how Alcuin's problem models this form of logic in a concrete way.

A genre of puzzle that plays on the same kind of logic, albeit in a slightly different way, is the "drawing out" puzzle, invented by the late American puzzle-maker, Martin Gardner (1914–2010). Below is a paraphrase:

In a box there are 20 billiard balls, 10 white and 10 black. They all feel the same. With a blindfold on, what is the least number you must draw in order to get a pair of balls that match in color (two white or two black)?

Here is the breakdown:

- Suppose the first ball we draw is white. If we are lucky, the next ball will also be white, and it's game over. The same reasoning applies to drawing two black balls in a row.
- But we cannot assume this lucky or best-case scenario, because the puzzle tells us that we *must* get a matching pair, luck notwithstanding. So, on the contrary, we assume a worst-case scenario, that is, we must assume that the first two draws produce balls of different colors.
- Let us suppose we draw a white ball first. Then, under the worst-case scenario, we will draw a black ball next. Thus, after two draws, we will have drawn one white and one black ball from the box.
- Obviously, we could have drawn a black ball first and a white one second, under the same scenario. The end result would have been the same: one white and one black ball.
- Now, no matter what the color is of the third ball we draw out, it will match the color of one of the two we had already pulled out, since there are only white and black balls in the box.

- If it is white, we will have two white balls; if it is black, we will have two black balls.
- So, the least number of balls we will need to draw from the box in order to be absolutely sure to get a pair of balls that matches is three.

What happens if we increase the number of colors? If there are 10 white, 10 black, and 10 green billiard balls in the box, how many draws are necessary to ensure a matching pair? Again, according to a worst-case scenario we will need to draw four balls, since we could draw a black, a white, and a green one in some order first, under this scenario. However, the next one we draw out will match one of these. If we continue in this way, we will come to a general principle: the number of draws required to ensure a match is one more than the number of colors—that is, if there are n colors then the number of draws is $(n + 1)$.

Now, what happens if we change the number of balls?

In a box there are 21 balls, 2 white, 2 black, 14 green, and 3 red. With a blindfold on, what is the least number you must draw in order to get a pair of balls that matches in color?

Because of the worst-case scenario requirement, the different number of balls does not alter the general rule.

- We could draw a white, a black, a green, and a red ball in some order first, under the worst-case scenario.
- But the next draw after that would be a ball in one of these colors.
- So, there is no change to the rule.

If socks or gloves are involved, rather than balls, then the reasoning changes somewhat, as the example below shows:

If there are six pairs of black gloves and six pairs of white gloves in a drawer, all mixed up, what is the least number of draws that are required in order to guarantee a matching pair of black or white gloves?

Because some gloves fit on the right hand and some on the left hand, one might pick all 12 left-handed gloves, of both colors, as a worst-case scenario. However, the thirteenth glove will be a right-handed one and also match one of the previous twelve in color. Thus, in this case, 13 gloves have to be drawn out in order to ensure that we get a pair of matching gloves. The same reasoning applies if we first draw out 12 right-handed gloves.

Exponents

Alcuin's problem of the sow and the pigpen (number 41) involves exponents. As mentioned, the *exponent* of a number tells us how many times the base must be used in a multiplication. It is worth going through some of the basic properties of numbers with exponents here for the sake of illustration, since these are fundamental to all of mathematics, as Alcuin seems to have intuited. Exponents are used not only with integers (positive and negative), but with fractions, decimals, and other types of numbers:

$$-4^3 = -4 \times -4 \times -4 = -64$$
$$1^5 = 1 \times 1 \times 1 \times 1 \times 1 = 1$$
$$.03^4 = .03 \times .03 \times .03 \times .03 = 0.00000081$$

Multiplying two exponential numbers with the same base is equivalent to adding their exponents: $3^4 \times 3^5 = 3^{4+5} = 3^9$. To see why this is so, consider what 3^4 times 3^5 tells us to do—namely to multiply 3 four times by 3 five times:

3^4	$\times$	3^5	$=$	3^9
$\downarrow$		$\downarrow$		$\downarrow$
$(3 \times 3 \times 3 \times 3)$	$\times$	$(3 \times 3 \times 3 \times 3 \times 3)$	$=$	$(3 \times 3 \times 3 \times 3 \times 3 \times 3 \times 3 \times 3 \times 3)$

As can be seen, there are nine "3's" in the product, which is the same as adding up the two exponents $(4 + 5 = 9)$. In general, if "a" is any base, then:

$$a^n \times a^m \times a^p \times \ldots = a^{n+m+p+\ldots}$$

If the bases are different but the exponents are the same, then the rule is as follows:

$$(a^n)(b^n) = (ab)^n$$

Consider a concrete example:

$$(5^3)(4^3) = (5 \times 4)^3 = 20^3 = 8000$$

To see why this rule applies, consider what it really means:

$$5^3 \times 4^3 = (5 \times 5 \times 5) \times (4 \times 4 \times 4)$$

Let us rearrange these in pairs as shown below:

$$
\begin{array}{ccccccc}
5^3 & \times & 4^3 & & \textit{In pairs} & & (ab)^n \\
\downarrow & & \downarrow & & \downarrow & & \downarrow \\
(5 \times 5 \times 5) & \times & (4 \times 4 \times 4) & = & (5 \times 4)(5 \times 4)(5 \times 4) & = & (5 \times 4)^3
\end{array}
$$

Dividing exponential numbers is equivalent to subtracting their exponents:

$$3^5 \div 3^3 = 3^{5-3} = 3^2$$

This is so because the exponent in "3^5" as the numerator in the division tells us to multiply "3" five times, and the exponent in "3^3" as the denominator tells us to multiply "3" three times:

$$\frac{3^5}{3^3} = \frac{(3 \times 3 \times 3 \times 3 \times 3)}{(3 \times 3 \times 3)}$$

Canceling in the usual way, we get:

$$\frac{3^5}{3^3} = \frac{\cancel{(3 \times 3 \times 3)} \times 3 \times 3}{\cancel{(3 \times 3 \times 3)}}$$

Since the result, 3×3, is 3^2, we can see that it is the same as subtracting the two exponents ($5 - 3 = 2$). In general:

$$a^n \div a^m = a^{n-m}$$

Consider one last rule, namely that any number raised to the power of "0" is equal to "1":

$$2^0 = 1$$
$$4^0 = 1$$
$$13^0 = 1$$

Consider the following example:

$$3^5 \div 3^5 = 1$$

This can be written out in full as follows:

$$\frac{3^5}{3^5} = \frac{\cancel{3 \times 3 \times 3 \times 3 \times 3}}{\cancel{3 \times 3 \times 3 \times 3 \times 3}}$$

The result is "1," after cancellation. We know from a rule above that $3^5 \div 3^5 = 3^{5-5} = 3^0$, and since $3^5 \div 3^5 = 1$, we conclude that $3^0 = 1$. In general:

$$n^0 = 1 \text{ (where "}n\text{ is any number)}$$

There are other rules, which need not concern us here. It is remarkable to consider that right after the exponent notation was introduced into mathematics to simplify a certain type of multiplication problem, it took on a life of its own, leading to an expansion of mathematics itself. This episode in the history of mathematics brings out the power of notation itself as a means not only to represent mathematical ideas compactly but also to reveal properties that may be hidden in them.

Logarithms

The *logarithm* is the exponent itself to which a base must be raised to produce a given number. For example: How many 2's must be multiplied to get 8? The answer is 3, which is the exponent for 2 used as a base:

$$2 \times 2 \times 2 = 8$$
$$2^3 = 8$$

So:

$$\log_2 8 = 3: \text{ The logarithm of 8 with base 2 is 3.}$$

In general:

If $m^a = n$, then

$\log_m n = a$: The logarithm of n with base m is a.

The most common base used for logarithms is 10, called the *common logarithm*:

$$10^3 = 1000: \text{ therefore, } \log_{10} 1000 = 3$$
$$10^{.69897} = 5 \text{ (rounded off): therefore, } \log_{10} 5 = .69897$$
$$10^{1.8633} = 73 \text{ (rounded off): therefore, } \log_{10} 73 = 1.8633$$

Logarithms were invented by Scottish mathematician John Napier (1550–1617) in his 1614 book, *Mirifici Logarithmorum Canonis Descriptio* ("Description of the Marvelous Canons of Logarithms"), in which he provided tables of logarithms. Today, logarithms are computed instantly by electronic computers. To get a sense of the utility of logarithms, suppose we wanted to calculate in which generation we had 1024 ancestors. The answer can be reached quickly as follows: $\log_2 1024 = 10$, which tells us that we have this number in the tenth generation before us. Let us go through the reasoning for the sake of illustration. Note that the base in this case is 2:

- We have two parents, so we have two ancestors in the first generation. This can be expressed as $2^1 = 2$ and thus $\log_2 2 = 1$ (noting that "1" stands for the first generation).
- Each of our parents has two parents, and so there are $2 \times 2 = 2^2 = 4$ ancestors in the second generation, shown with the logarithm as follows: $\log_2 4 = 2$ (noting again that "2" stands for the second generation).
- Each of the four grandparents had two parents, and so there were $4 \times 2 = 2 \times 2 \times 2 = 2^3 = 8$ ancestors in the third generation: $\log_2 8 = 3$ (with the "3" standing for the third generation).
- The calculation continues with the same method.
- So, in which generation do we have 1024 ancestors? We can rephrase the question as follows: For which exponent "n" is it true that $2^n = 1024$?
- Using logarithmic notation, the answer is 10, $\log_2 1024 = 10$, because $2^{10} = 1024$. This tells us that in the tenth previous generation we had 1024 ancestors.

Logarithms are the reverse of exponentiation, so computations with logarithms are governed by the same type of rules. Below are a few examples using the common logarithm, $\log_{10}$:

- *Product Rule:*
$\log_{10}(m \times n) = \log_{10} m + \log_{10} n$
Example
$\log_{10}(2 \times 3) = \log_{10} 2 + \log_{10} 3 = 0.3010 + 0.4771 = 0.7781$

- *Quotient Rule:*
$\log_{10}(m/n) = \log_{10} m - \log_{10} n$
Example
$\log_{10}(5/7) = \log_{10} 5 - \log_{10} 7 = 0.6989 - 0.8450 = -0.1461$

- *Power Rule:*
$\log_{10}(m^n) = n \times \log_{10} m$
Example
$\log_{10}(7^5) = 5 \times \log_{10} 7 = 5 \times 0.8450 = 4.225$

- *Zero Rule:*
$\log_{10}(1) = 0$, since $10^0 = 1$

Logarithms have been useful to scientists because they simplify long, tedious calculations, and because they have been found to explain a host of natural phenomena.

There is one other type of widely used logarithm in mathematics and science, represented as "ln (n)," called the *natural logarithm*. This is the logarithm of a number using the base "e" (known as "Euler's number," even though it was known before Euler), which has the value 2.71828.... Below are a few values for ln (n), compared to common logarithms:

Table 2.1 Values for ln (n), compared to common logarithms.

n	$\log_{10}(n)$	$\ln(n)$
1	0	0
3	0.4771	1.0986
15	1.1760	2.7080
21	1.3222	3.0445
48	1.6812	3.8712
123	2.0899	4.8121
. . .	. . .	. . .

The natural logarithm appears in equations describing growth and change; it surfaces in formulas for curves; it crops up frequently in probability theory; it surfaces in formulas for calculating compound interest; among many other areas of mathematics, science, and everyday life. A commonly used expression for determining the value of "e" is $(1 + 1/n)^n$ as $n \to \infty$, devised by the Swiss mathematician Jacob Bernoulli in the late seventeenth century (Bernoulli 1690). This says that "e" is determined to greater degrees of accuracy by increasing values of n as it goes to infinity. Below are a few examples:

Table 2.2 Examples of "e" determined to greater degrees of accuracy by increasing values of n.

n	$(1 + 1/n)^n$	e
1	$(1 + 1/1)^1$	2.0
2	$(1 + 1/2)^2$	2.25
5	$(1 + 1/5)^5$	2.4883 . . .
10	$(1 + 1/10)^{10}$	2.5937 . . .
100	$(1 + 1/100)^{100}$	2.7048 . . .
10,000	$(1 + 1/10,000)^{10,000}$	2.7181 . . .
. . .	. . .	. . .

Impossibility and Provability

The problem of the pigs (number 43) enfolds a fundamental issue in mathematics—impossibility—or the fact that a particular problem cannot be solved as described. Consider the following similar problem:

Find five consecutive odd numbers that add up to 64.

Let us start by considering the sum of the first five consecutive odd numbers:

$$1 + 3 + 5 + 7 + 9 = 25$$

If we continue adding sets of five consecutive odd numbers we will find that the sum always turns out to be odd. It would seem, therefore, that it is impossible for five consecutive odd numbers to add up to an even sum.

Box 2.2 Representing Even and Odd Numbers

An even number is represented with the expression $2n$, with $n = \{\pm 1, \pm 2, \pm 3, \pm 4, \pm 5, \pm 6, \ldots \}$, which simply states that any number, n, multiplied by 2 results in an even number. Examples:

If $n = 3, 2n = 2 \times 3 = 6$ (positive even integer)

If $n = -3, 2n = 2 \times -3 = -6$ (negative even integer)

If $n = 25, 2n = 2 \times 25 = 50$ (positive even integer)

...

For an odd number the expression is either $(2n + 1)$ or $(2n - 1)$, which represents the fact that any number before or after an even number will be odd in the normal order: $\{\underline{1}, 2, \underline{3}, 4, \underline{5}, 6, \underline{7}, 8, \underline{9}, \ldots \}$. Examples:

If $n = 3, (2n + 1) = (2 \times 3 + 1) = 6 + 1 = 7$ (positive odd integer)

If $n = -3, (2n + 1) = (2 \times -3 + 1) = -6 + 1 = -5$ (negative odd integer)

If $n = 25, (2n + 1) = (2 \times 25 + 1) = 50 + 1 = 51$ (positive odd integer)

Now, five consecutive odd numbers can be represented with $(2n + 1)$, $(2n + 3)$, $(2n + 5)$, $(2n + 7)$, and $(2n + 9)$. Adding these up yields the following result:

$$(2n + 1) + (2n + 3) + (2n + 5) + (2n + 7) + (2n + 9) = (10n + 25)$$

Consider the expression $(10n + 25)$. The term $10n$ is a number ending in 0 because any number, n, multiplied by 10, will invariably produce a number ending in 0: $1 \times 10 = 10$, $2 \times 10 = 20$, $15 \times 10 = 150$, and so on. The second term, 25, when added to the previous one ending in 0, will result in a number ending with the digit 5: $10 + 25 = 35$, $20 + 25 = 45$, $150 + 25 = 175$, and so on. So, the expression $(10n + 25)$ represents an odd digit ending in 5, no matter what n is.

The ancient Greek mathematicians grappled constantly with the concept of impossibility, wondering why, for example, it was seemingly impossible to trisect an angle with compass and ruler, given that angle bisection was such a simple procedure. For centuries after, mathematicians attempted trisection with compass and ruler, but always to no avail. The formal proof had to await the nineteenth century, published by mathematician Pierre Laurent Wantzel (1814–1848) in 1837.

Lewis Carroll was fascinated by the implications of impossibility in mathematics. In *Through the Looking-Glass* (1872), he quipped about it as follows:

> Alice laughed. "There's no use trying," she said; "one can't believe impossible things."
> "I daresay you haven't had much practice," said the White Queen. "When I was your age, I always did it for half-an-hour a day. Why, sometimes I've believed as many as six impossible things before breakfast."

Related to impossibility is provability. That is to say, can every mathematical proposition be proved or disproved? In a famous 1931 paper, mathematician Kurt Gödel (1906–1978) showed that a mathematical system invariably contained a proposition within it that is "true" but "unprovable." His proof is complex and beyond the present scope. A good easy-to-follow explanation can be found in the book, *An Introduction to Gödel's Theorem* (1958), by Ernest Nagel and James R. Newman. Gödel's argument can be condensed as follows:

> Consider a mathematical system, T, in which: (a) no false statement is provable in it and (b) contains a statement "S" that asserts its own unprovability. S can be formulated simply as: "I am not provable in system T." What can we make of S? If it is false, then its opposite is true, namely, "I am provable in T," which means, of course, that S is provable in T. But this goes contrary to our assumption (a) that no false statement is provable in the system. Therefore, we conclude that S must be true, from which it follows that S is unprovable in T, as S asserts. Thus, S is true, but not provable in the system.

The implications of this proof are still reverberating. But mathematics marches on (pardon the cliché). As David Tall (2013: 246) comments, "Instead of trying to prove *all* theorems in an axiomatic system (which Gödel showed is not possible), professional mathematicians continue to use a formal presentation of mathematics to specify and prove many theorems that are amenable to the formalist paradigm."

In mathematics, something that seems to be true and thus provable in some way, but eludes proof, is called a conjecture. One of the most famous conjectures appeared in the eighteenth century, when, in a 1742 letter he wrote to Leonhard Euler, Prussian mathematician Christian Goldbach claimed that it seemed that every even integer greater than 2 was the sum of two primes:

$$4 = 2 + 2$$
$$6 = 3 + 3$$
$$8 = 3 + 5$$
$$10 = 3 + 7$$
$$...$$
$$100 = 47 + 53$$

Box 2.3 Prime Numbers

A *prime number* is an integer that has no divisors (factors) other than itself and 1. For example, 5 and 7 are primes because they can only be divided by themselves and 1. Composite numbers are those that can be broken down into prime factors. For example, $48 = 2 \times 2 \times 2 \times 2 \times 3 = 2^4 \times 3$, showing that the prime factors of 48 are 2 and 3.

Each composite number is a product of one, and only one, distinct set of prime factors. This is called the *Fundamental Theorem of Arithmetic*. It was formulated by Euclid and proved by the German mathematician Carl Friedrich Gauss (1777–1855) two millennia later. Given any composite number, such as 14 or 50, the theorem states that it is decomposable into a unique set of prime factors:

$$14 = 2 \times 7 \text{ (factors: } 2, 7)$$
$$50 = 2 \times 5 \times 5 = 2 \times 5^2 \text{ (factors: } 2, 5)$$

In the same letter to Euler, Goldbach also conjectured that any number greater than 5 could be written as the sum of three primes:

$$6 = 2 + 2 + 2$$

$$8 = 2 + 3 + 3$$

$$7 = 2 + 2 + 3$$

$$9 = 3 + 3 + 3$$

$$10 = 2 + 3 + 5$$

$$11 = 3 + 3 + 5$$

...

Euler wrote back that he was certain that the conjectures were true, but that he was unable to prove it. Ever since, there have been many attempts to prove the Goldbach conjectures (for example, Helfgott 2013). But so far no definitive proof has been formulated. The point here is that the search for a proof or solution to a problem that seems intractable makes manifest how mathematical curiosity cannot be curtailed, once it has been stimulated. Maybe Alcuin put his Problem 43 in his text as a way to stimulate curiosity in his students. There are many stories of mathematicians becoming obsessed with finding solutions to intractable problems. This is a theme expounded interestingly in a graphic novel, *Logicomix: An Epic Search for Truth* (2009), by Apostolos Doxiadis, in which the personal struggles of famous mathematicians are portrayed as dovetailing with their almost-obsessive need to resolve issues in the field. Doxiadis had previously written the novel, *Uncle Petros and Goldbach's Conjecture* (2000), in which the quest to solve Goldbach's conjectures is depicted as the essence of mathematical curiosity.

Epilogue

The problems discussed in this chapter imply that Alcuin understood the power of abstract thinking as a means to grasp hidden truths. Some of the problems may have been sparked by an actual experience or observation related to a particular situation. Once solved we feel a sense of satisfaction, as well as having better understood the features of the real-world experience that undergirded the contents and structure of the problems.

The notion of *hypothesis* is key to understanding many aspects of mathematics. If the situation that was translated into a problem was reduced to a simple mechanical arithmetical activity, it would remain meaningless as a tool for understanding some principle inherent in it. Hypothesis thinking transforms a concrete situation into a theoretical one. It turns everyday information into abstract representation, attenuating, or even eliminating the need for mechanical trial and error. All puzzles are, in a sense, experiments in hypothesis thinking.

Explorations

The 10 explorations here provide different perspectives of the various themes and methods covered in this chapter. Some of these are versions of classic puzzles in recreational mathematics.

1. *A Counting Problem*

There are between 50 and 60 pigs in a pigpen. If we count them three at a time, we will find that there are two left over. If, however, we count them five at a time, we will find that there are four left over. How many pigs are there in the pen?

[This is a type of problem found commonly in school texts and even in recreational mathematics collections. It has been adapted here in imitation of the content and style of Alcuin's problems.]

2. *Balance Scale Puzzle*

There are six billiard balls, one of which weighs less than the other five. Otherwise, they all look and feel identically the same. How can the one that weighs less be identified on a balance scale with only two weighings?

[This is a modified version of a classic puzzle in recreational mathematics, devised by Martin Gardner (1979). It is included here because it constitutes an example of the use of hypothetical thinking in mathematics—that is, trying out certain combinations of balls to weigh hypothetically to see what they yield.]

3. *Bachet's Weighing Puzzle*

What is the least number of weights that can be used on a scale to weigh any whole number of pounds of sugar from 1 to 40 inclusive, if the weights can be placed on either of the scale pans?

Box 2.4 Claude Gaspar Bachet de Méziriac (1581–1683

The puzzle above was devised by French mathematician and poet Claude Bachet in his 1612 collection of puzzles, titled *Problèmes plaisans et délectables qui se font par les nombres* ("Pleasant and Delectable

Problems with Numbers"), which contains the puzzle above. He also published a Latin translation of the *Arithmetica* of Diophantus— the very translation in which Pierre de Fermat wrote a note in the margin that became *Fermat's Last Theorem* (discussed subsequently). Bachet also developed a technique for solving indeterminate equations using continued fractions, as well as other important mathematical ideas. Bachet's puzzle is the likely inspiration for subsequent weighing puzzles, such as Exploration 2 above.

---

4. *A Puzzle by Lewis Carroll*

A bag contains one counter, known to be either white or black. A white counter is put in, the bag shaken, and a counter drawn out (with a blindfold on), which proves to be white. What is now the chance of drawing a white counter on a second draw?

Box 2.5 Lewis Carroll (1832–1898)

This puzzle was devised by Lewis Carroll, pseudonym of Charles Lutwidge Dodgson, who was a creator of ingenious puzzles, even though he is best known for writing widely read children's novels. Carroll was a brilliant mathematician, who taught mathematics to undergraduates at Christ Church, Oxford for many years. In the preface to his 1893 puzzle book, titled *Pillow Problems Thought Out During Wakeful Hours*, Carroll stated that he chose this title because he wanted the puzzles to "cure" the reader's insomnia, from which Carroll suffered, paradoxically suggesting that his problems would keep people up at night trying to figure them out. He also wrote that the problems would also prevent distress, as well as avert "unholy thoughts which torture with their hateful presence the fancy that would fain be pure."

---

5. *A Take on Alcuin's Problem 43*

A farmer wanted to sell his 228 pigs on five consecutive days. Being mathematically inclined, the farmer decided to sell them in groups consisting of consecutive odd numbers of pigs per day. How did he do it?

6. *Hilbert's Infinite Hotel Problem*

Imagine a hotel with an infinite number of rooms, and all the rooms occupied. To the hotel comes a new guest who asks for a room. The proprietor says that he can easily accommodate the guest. How does he do it?

[This problem and the next one were formulated by mathematician David Hilbert (1862–1943) in 1924, and treated by George Gamow in his 1947 book, *One, Two, Three . . . Infinity.*]

7. *Another Infinite Hotel Problem*

Imagine a hotel with an infinite number of rooms, all taken up, and an infinite number of new guests who come in and ask for rooms. Again the proprietor is able to accommodate them. How so?

8. *A Take on Alcuin's Problem 41*

A farmer notices that the number of animals in his farm has spiraled over a span of time. Each of the original three chickens gave birth to three chickens of their own. The original four pigs did the same, each one giving birth to four pigs of their own. The original five cows also did the same, each one giving birth to five cows of their own. Being mathematically inclined, he noticed that if he added two more animals he would then have a number that is a perfect square. What number is it?

[As in Alcuin's problem, the ways in which the offspring were produced are implausible and must be ignored for the sake of the problem.]

9. *Census Taker Conundrum*

A census taker asks a resident about her children. The resident says, "I have three children. The product of their ages is thirty-six. The sum of their ages is the number on the gate to my house." The census taker says that this is not enough information. However, after the woman says, "My oldest child has the measles," the census taker figures it out. What are the children's ages?

[This problem has appeared in different puzzle anthologies. According to Greenblatt (1965), it likely originated around 1940, during World War II, on the MIT campus.]

10. *Drawing Out Problem*

In a box, there are 100 balls, 20 white, 19 black, 18 green, 15 blue, and 1 yellow. With a blindfold on, what is the least number you must draw out in order to get a pair of balls that match in color?

[This is a variant of the problem discussed in the Annotations.]

Cited Works and Further Reading

Archimedes (c. 216 BCE). *The Sand Reckoner.* Internet Archive, https://archive.org/details/sandreckoner00brad.
Bachet, Claude-Gaspar (1612). *Problèmes plaisans et délectables qui se font par les nombres.* Lyon: Gauthier-Villars.

Bernoulli, Jacob (1690). Quæstiones nonnullæ de usuris, cum solutione problematis de sorte alearum, propositi in Ephem. Gall. A. 1685 ("Some Questions about Interest, with a Solution of a Problem about Games of Chance, Proposed in the *Journal des Savants* (Ephemerides Eruditorum Gallicanæ), in the Year (anno) 1685), *Acta Eruditorum*, pp. 219–223.

Burkholder, Peter (1993). Alcuin of York's *Propositiones ad acuendos juvenes*: Introduction, Commentary & Translation. *History of Science & Technology Bulletin*, Vol. 1, number2.

Cantor, Georg (1874). Über eine Eigneschaft des Inbegriffes aller reelen algebraischen Zahlen. *Journal für die Reine und Angewandte Mathematik* 77: 258–262.

Carroll, Lewis (1872). *Through the Looking-Glass and What Alice Found there*. London: Macmillan.

Carroll, Lewis (1893). *Pillow Problems Thought Out During Wakeful Hours*. London: Macmillan.

Collingwood, Stuart Dodgson (1899). *The Lewis Carroll Picture Book*. London: T. Fisher Unwin.

Doxiadis, Apostolos (2000). *Uncle Petros and Goldbach's Conjecture*. New York: Bloomsbury.

Doxiadis, Apostolos (2009). *Logicomix: An Epic Search for Truth*. New York: Bloomsbury.

Euclid (c. 300 BCE). *The Thirteen Books of Euclid's Elements*, 3 Volumes. New York: Dover, 1956.

Euler, Leonhard (1736). *Mechanica, sive motus scientia analytice exposita*. Petropoli: Typographia Academiae Scientiarum.

Galilei, Galileo (1638). *Dialogue Concerning the Two Chief World Systems*, trans. by Stillman Drake. New York: Modern Library, 2001.

Gamow, George (1947). *One, Two, Three...Infinity*. New York: Dover.

Gardner, Martin (1979). *A Mathematical Circus*. Washington: Mathematical Association of America Press.

Gödel, Kurt (1931). Über formal unentscheidbare Sätze der Principia Mathematica und verwandter Systeme, Teil I. *Monatshefte für Mathematik und Physik* 38: 173–189.

Greenblatt, M. H. (1965). *Mathematical Entertainments*. New York: Crowell.

Helfgott, Harald A. (2013). The Ternary Goldbach Conjecture Is True. arXiv:1312.7748.

Levin, Leonid (1971). Some Theorems on the Algorithmic Approach to Probability Theory and Information Theory. *Annals of Pure and Applied Logic* 162 (2010): 224–235.

Lützen, Jesper (2023). *A History of Mathematical Impossibility*. Oxford: Oxford University Press.

Maor, Eli (1994). *e: The Story of a Number*. Princeton: Princeton University Press.

Nagel, Earnest and Newman, James R. (1958). *Gödel's Proof*. New York: New York University Press.

Napier, John (1614). *Mirifici logarithmorum canonis constructio*. Edinburgh: Blackwood and Sons, 1889.

Petkovic, Miodrag S. (2009). *Famous Puzzles of Great Mathematicians*. Providence, RI: American Mathematical Society.

Tall, David (2013). *How Humans Learn to Think Mathematically*. Cambridge: Cambridge University Press.

Wantzel, Pierre Laurent (1837). Recherches sur les moyens de reconnaître si un Problème de Géométrie peut se résoudre avec la règle et le compass. *Journal de Mathématiques Pures et Appliquées* 2: 366–372.

3
Geometry

Prologue

The term *geometry* comes from the Greek *geo* "earth" and *metrein* "to measure." It describes perfectly the activities of the ancient builders and architects, who were concerned with such problems as measuring the size of fields and determining how to make accurate right angles in the construction of buildings. It was in the seventh century BCE, when Thales of Miletus (c. 623–545 BCE) introduced the method of proof as a means to establish inherent principles of geometric structure and of the mathematical logic behind them. By 300 BCE, Euclid incorporated and developed this method as the central feature of mathematical thinking in his *Elements*.

Ever since, Euclidean geometry has been considered a core subject at school. The importance of geometry in antiquity is borne out by the inscription over the entrance into Plato's academy: "Let no one who is unacquainted with geometry enter here." The theoretical and practical significance that was assigned to geometry in antiquity continued to shape education (and mathematics itself) in the medieval ages, as can be seen in texts such as the one by Alcuin.

The Problems

There are six problems in the *Propositiones* dealing specifically with geometry—numbers 21, 22, 23, 24, 25, and 26. Several of them seem to be "off" mathematically, perhaps because Alcuin may have been challenging his readers to think "outside the box," as the expression goes today.

21. *Problem of Sheep in a Field*

There is a field that is 200 feet long and 100 feet wide. A farmer wants to put sheep in it as follows: each sheep should have an area 5 feet long and 4 feet wide. How many sheep can be put in such a field?

22. *Problem of an Irregular Field*

There is an irregular field which is 100 feet on each side, 50 feet on one front, 60 feet in the middle, and 50 feet on the other front. How many square feet does this field enclose?

23. *Problem of the Quadrangular Field*

There is a field which is 30 feet on one side, 32 feet on another, 34 feet in the front, and 32 feet on the remaining side. How many square feet are contained in such a field?

24. *Problem of the Triangular Field*

There is a field which is 30 feet on one side, 30 feet on another, and 18 feet in the front. What is the area of such a field?

25. *Problem of the Round Field*

There is a round field, 400 feet in circumference. How many square feet will its area be?

26. *Problem of a Dog Chasing a Hare*

There is a field which is 150 feet long. At one end stood a dog, at the other, a hare. The dog chases the hare, advancing 9 feet per leap, while at the same time as the dog, the hare makes a 7-feet leap. How many feet did the dog travel and how many leaps did it take in pursuing the fleeing hare until it caught the hare?

Solutions

21. *Answer:* 1000 sheep

Solution

The problem can be envisaged as designing a grid pattern inside the field consisting of small rectangles of the same size, each of which measures the area allotted to each sheep—namely 5 feet by 4 feet. Figuring out how many of these small squares there are will then reveal the number of sheep. Below is a breakdown of the solution:

- The dimensions of the field are 200 by 100.
- The area of the field is, therefore: $200 \times 100 = 20{,}000$ square feet.
- The area allotted to each sheep is: $5 \times 4 = 20$ square feet.
- To find out how many of these there are in the field, we divide 20,000 by 20, which is 1000.
- That is how many sheep can be placed in the field according to Alcuin's specifications.

22. *Answer:* 5333 square feet (approximately)

Solution

It is not clear what shape Alcuin had in mind for his field. In fact, his own answer is dubious, as Hadley and Singmaster (1992) point out in their commentary on this problem. Alcuin uses an approximation method, essentially assuming that the area is that of a rectangle. His solution can be elaborated as follows:

- Alcuin seems to consider the 50, 50, and 60 feet he describes as measures that can be united together: 50 + 50 + 60.
- He then assumes that 1/3 of this sum would be one of the units to be used in calculating the area, say, the length of the field: 1/3 (50 + 50 + 60). It is not clear why this is so, however.
- The other measure he describes is 100 feet, and this can be considered to be the width of the field.
- Therefore, the area, which is determined by multiplying the length times the width would be: 1/3 (50 + 50 + 60) feet × 100 feet = 5333 square feet (approximately).
- We will leave it at that, since it is almost impossible to envision how Alcuin drew the field in his mind or how he came up with his calculations.

23. *Answer:* 1020 square feet

Solution

Again, it is hard to imagine what Alcuin meant by this problem, given his designation of the field as quadrangular. As O'Connor and Robertson (2012) aptly comment, "This question can't be solved since giving the lengths of the sides of the field does not determine it." Alcuin gives the solution of 1020 square feet. One way to envisage the problem is that Alcuin gave different dimensions to confuse the students, or to get them to come up with different possibilities. Of these, only one matches Alcuin's answer—the first one below:

- If we take the dimensions of the field to be 30 × 34, then 1020 square feet is the area.
- If we take them to be 32 × 32, then 1024 square feet is the area.
- If we take them to be 30 × 32, then 960 square feet is the area.
- If we take them to be 32 × 34, then 1088 square feet is the area.

24. *Answer:* 270 (according to Alcuin), 257.58 square feet (actual area)

Solution

The field has the shape of an isosceles triangle. The two equal sides are 30 feet in length and the base is 18 feet. This can be shown with a diagram (Figure 3.1).

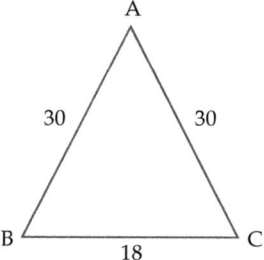

Figure 3.1 Diagram for Alcuin's Triangular Field

If we drop a perpendicular line from the vertex (A) of the triangle to its base (BC) it will bisect the base—an established fact about isosceles triangles (Figure 3.2).

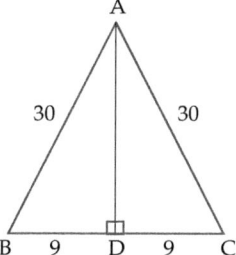

Figure 3.2 Perpendicular Drawn to the Triangle

The perpendicular thus produces two congruent right triangles. To calculate the area of either one, we multiply half the base by the perpendicular height. The height is AD. To figure out its length, we can use the Pythagorean theorem, which states that the square on the hypotenuse (*c*) of a right-angled triangle is equal to the sum of the squares on the other two sides (*a*, *b*) (Figure 3.3).

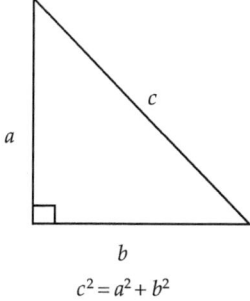

$$c^2 = a^2 + b^2$$

Figure 3.3 Pythagorean Theorem

We can take either one of the two right triangles, say, ADC (Figure 3.2). In the triangle, DC = 9 and AC = 30, which is the hypotenuse. Using the Pythagorean theorem:

$$c^2 = a^2 + b^2$$

$$30^2 = 9^2 + AD^2$$

$$AD^2 = 30^2 - 9^2$$

$$AD^2 = 900 - 81 = 819$$

$$AD = \sqrt{819} = 28.62 \text{ (rounded off)}$$

AD is the height of the triangle; so $h = 28.62$. We also know the base of the triangle, $b = 9 + 9 = 18$. Thus, the area of the triangular field is as follows:

$$\tfrac{1}{2} (b) (h)$$

$$\tfrac{1}{2} (18) (28.62) = 257.58 \text{ square feet}$$

Alcuin computes it to be 270 square feet. Either his solution is a miscalculation, or it was changed by copyists of the manuscript. In an era before the printing press, copying by hand (over and over) was bound to lead to errors or unwarranted changes.

25. *Answer*: 10,000 square feet (according to Alcuin), 12,743.60 square feet (actual area)

Solution

The field is a circle. The area of a circle, A, is given by the formula $A = \pi r^2$, with $\pi = 3.14 \ldots$ and r = the radius. We will need to determine what r is before computing the area. Alcuin gives the circumference (C) as 400. Since $C = 2\pi r$, therefore:

$$C = 2\pi r$$

$$2\pi r = 400$$

$$r = 400/2\pi$$

$$r = 63.69$$

We can now calculate the area:

$$A = \pi r^2$$

$$A = \pi (63.69)^2 = 12,743.60 \text{ square feet.}$$

Alcuin comes up with the answer of 10,000, by using the approximation of $\pi = 4$, for some reason. Perhaps, as with his other slightly anomalous solutions, this may have been the result of some copying error.

26. *Answer:* 675 feet, 75 leaps

Solution

The initial state can be represented with a diagram (Figure 3.4), showing the dog (D) and the hare (H) separated by 150 feet, with E representing the end point where the dog eventually catches the hare.

Figure 3.4 Initial Positions of the Dog and Hare

Let the amount of time taken for each leap by the dog or the hare be x units, since the problem says that the time taken was the same for both. The dog, whose rate is 9 feet per leap, will thus cover $9x$ feet in total, from its initial position to the end point E, where it catches the hare. In the same time frame, the hare, whose rate is 7 feet per leap, will cover $7x$ feet from its own starting point to E, where it is caught by the dog (Figure 3.5).

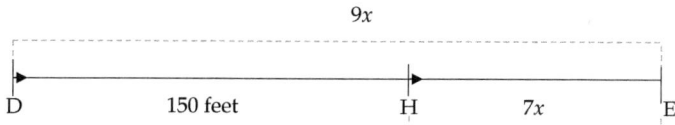

Figure 3.5 Distance Covered by the Dog and the Hare to E

As can be seen in the figure, the total distance covered by the dog, $9x$, is equal to the initial 150 feet between itself and the hare and the distance, $7x$, covered by the hare to E.

$$9x = 150 + 7x$$
$$9x - 7x = 150$$
$$2x = 150$$
$$x = 75 \text{ time units (one unit = one leap)}$$

This means that the dog took 75 leaps. From this we can calculate the dog's distance as $9x = 9 \times 75 = 675$ feet.

Annotations

Alcuin's Problem 21 requires imagining a field as a grid, with its length calibrated in units of 5 feet and its width in units of 4 feet, allowing for a tessellation of the entire field with 5-by-4 feet little squares. Problems 22 and 23, on the other hand, require some truly lateral thinking, since it is almost impossible to infer what kind of shapes Alcuin had in mind. Could they have been problems intended to get his readers to explore geometrical notions with their own imaginations? The problem of the dog and the hare brings out the importance of devising an appropriate diagram to represent a geometry problem. In this case, the diagram allows us to envision the relations among the leaps covered by the dog and those by the hare in outline form.

In effect, these problems constitute, together, a mini-treatise in geometrical thinking. They also show how such thinking has practical applications, even if the situations described are themselves somewhat artificial.

Quadrangular Figures

Several of Alcuin's problems deal with rectangles and squares, which were seen in antiquity to be related to arithmetical notions, starting with the Pythagoreans. The numbers were called "figurate." As an example, consider squares of increasing size, corresponding to the squares of the integers in order (Figure 3.6).

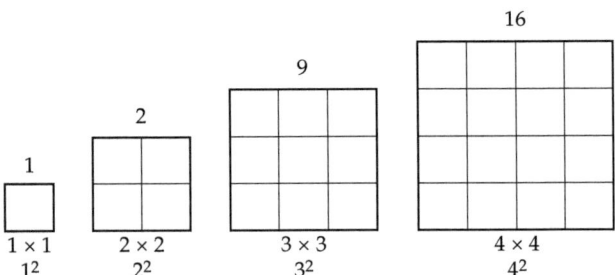

$$1 \times 1 \qquad 2 \times 2 \qquad 3 \times 3 \qquad 4 \times 4$$
$$1^2 \qquad\quad 2^2 \qquad\quad 3^2 \qquad\quad 4^2$$

Figure 3.6 Square Numbers

Each successive square is constructed as follows:

- The 1×1 square is the first, unit square.
- The 2×2 square is constructed by adding *three* unit squares to the first square: $1 + 3 = 4$ square units.
- The 3×3 square is constructed by adding *five* unit squares to the previous 2×2 square: $1 + 3 + 5 = 9$ square units.

- The 4 × 4 square is constructed by adding *seven* unit squares to the previous 3 × 3 square: $1 + 3 + 5 + 7 = 16$ square units.
- And so on ad infinitum.

This method of construction clearly bears within it a series defined by the sum of the odd numbers in order:

$$1 = 1 \times 1 = 1^2 = \boxed{1}$$

$$4 = 2 \times 2 = 2^2 = 1 + 3$$

$$9 = 3 \times 3 = 3^2 = 1 + 3 + 5$$

$$16 = 4 \times 4 = 4^2 = 1 + 3 + 5 + 7$$

$$25 = 5 \times 5 = 5^2 = 1 + 3 + 5 + 7 + 9$$

...   ↑

Series: $\{1 + 3 + 5 + 7 + 9 + ...\}$

Findings such as this showed how geometry and arithmetic were intrinsically interrelated. The ancients also dissected and rearranged quadrangular figures to see what mathematical insights these might yield. The latter have also been a part of recreational mathematics from at least the eighteenth century when a 1774 book of puzzles titled *Rational Recreations* by William Hopper included what has come to be called the "chessboard paradox," which is the prototype for such puzzles (Fredrickson 2003).

We start by dividing a square piece of paper into 64 equal little squares. The result is an 8 × 8 square (Figure 3.7).

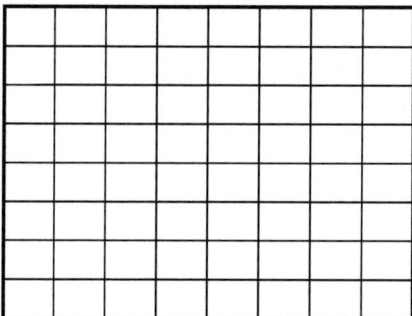

Figure 3.7 Chessboard Figure

Next, we draw two trapezoids (figures 1 and 2) and two triangles (figures 3 and 4) on the chessboard figure as shown in Figure 3.8. A trapezoid is a four-sided plane figure with two opposite sides parallel to each other.

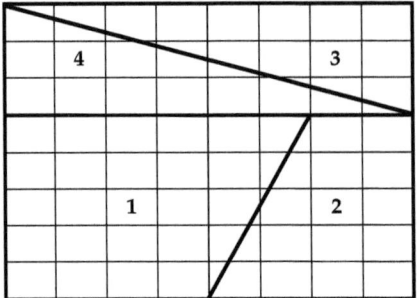

Figure 3.8 Figures Drawn on the Chessboard

Now, we cut these figures out and then rearrange them into a rectangle as shown in Figure 3.9.

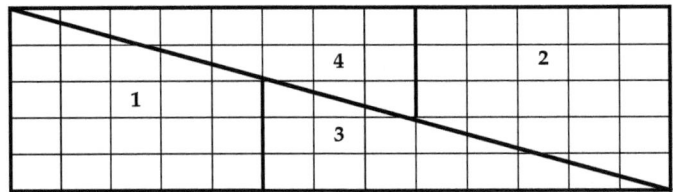

Figure 3.9 Rearrangement of Pieces

Let us count the small squares in the rectangle. There are now 5 × 13, or 65. But that is one more than the 64 that were in the original chessboard square we used to design the figures, cut them out, and then put them together to make the rectangle. How did the extra small square get in there? The truth is that the four figures do not really fit together perfectly but leave a barely visible gap around the diagonal in the shape of a tiny quadrilateral, thus producing the extra unit (Figure 3.10).

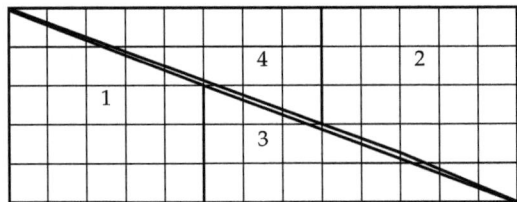

Figure 3.10 Hidden Tiny Quadrilateral

There are now many puzzles of this type in recreational mathematics. However, the prototype for this kind of puzzle is Archimedes' *loculus*

(Darling 2004, Netz and Noel 2007). It consisted of a square in which 14 geometric shapes were cut out. The main objective was to scramble the shapes and then reassemble them to reconstruct the original square (Netz and Noel 2007). In the tenth century, Arabic mathematicians used geometric dissections in their commentaries on Euclid's *Elements*. This topic will be taken up again in Chapter 10.

The Triangle

Alcuin's Problem 24 deals with the triangle, which, like the square, was seen as enfolding number properties in antiquity, starting again with the Pythagoreans. Consider the triangular shapes in Figure 3.11, with each dot standing for an integer.

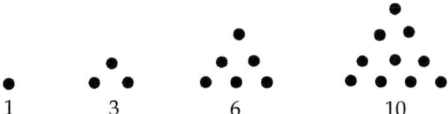

Figure 3.11 Triangular Numbers

The construction rule is as follows:

- The first triangular number (1) consists of one dot, corresponding to the unit number 1.
- The second triangular number (3) is constructed by adding two dots to the previous unit triangle, corresponding to the number 2, and hence $1 + 2 = 3$ dots.
- The third triangular number (6) is constructed by adding three dots to the previous triangle, corresponding to the number 3, and hence $1 + 2 + 3 = 6$ dots.
- The fourth triangular number (10) is constructed by adding four dots to the previous triangle, corresponding to the number 4, and hence $1 + 2 + 3 + 4 = 10$ dots.
- And so on ad infinitum.

This method of construction bears within it a series—the sum of the integers in order:

$$1 = 1$$
$$3 = 1 + 2$$
$$6 = 1 + 2 + 3$$
$$10 = 1 + 2 + 3 + 4$$
$$15 = 1 + 2 + 3 + 4 + 5$$
...

Series: $\{1 + 2 + 3 + ...\}$

The Pythagoreans came up with many connections between numbers and geometric figures, believing that this would allow them to understand the mysteries of the universe. As Aristotle wrote in his *Metaphysics* (350 BCE): "The Pythagoreans, who were the first to take up mathematics, not only advanced this subject, but saturated with it, they fancied that the principles of mathematics were the principles of all things."

To solve Alcuin's triangular field problem, the Pythagorean theorem was used. It states that in a right-angled triangle the square of the hypotenuse is equal to the sum of the squares of the other two sides. The ancient builders had discovered that stretching a rope into sides of 3, 4, and 5 units in length produced a right triangle, with 5 the longest side. The Pythagoreans were aware of this kind of discovery. Their aim was to prove that it revealed a general structural pattern, but did not leave any written evidence of a proof. As a result, we can only surmise how they carried it out. Proofs of the theorem have actually appeared in various areas of the ancient world, from India to China.

It is believed that the Pythagoreans used a dissection proof, such as the following one. We first draw a right-angled triangle with sides $\{a, b\}$ and hypotenuse $\{c\}$ (Figure 3.12).

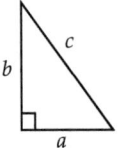

Figure 3.12 Right Triangle

Then, we construct a square with length $(a + b)$, the sum of the lengths of the two legs of the triangle. The result is equivalent to joining four copies of the triangle together (Figure 3.13).

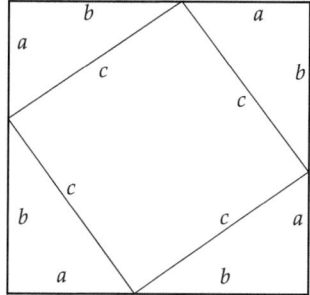

Figure 3.13 Square Constructed with Four Right Triangles

The diagram shows that the area of the square is $(a + b)^2$. The area of a triangle is half the base times the height. So, one triangle has an area of ½ ab, meaning that four triangles have an area of $4 \times ½\ ab = 2ab$. The area of the inner square is c^2. Therefore, the area of the big square, $(a + b)^2$, is equal to the sum of the areas of the four triangles, $2ab$, plus the area of the inner square, c^2 (note that quadratic equations will be discussed in Chapter 5).

$$(a + b)^2 = 2ab + c^2$$
$$a^2 + 2ab + b^2 = 2ab + c^2$$
$$c^2 = a^2 + 2ab + b^2 - 2ab$$
$$c^2 = a^2 + b^2$$

Theorem proved

There are actually hundreds of valid proofs for the theorem. The theorem also begs a question that has had repercussions for many areas of mathematics: If $c^2 = a^2 + b^2$, then are there infinite sets of triple integers that satisfy the general formula $c^n = a^n + b^n$, that is, $c^3 = a^3 + b^3$, $c^4 = a^4 + b^4$, and so on? French mathematician Pierre de Fermat (1607–1665) left a conundrum when he claimed that he had found a "simple proof" that $c^n = a^n + b^n$ had integral (whole number) solutions only if $n = 2$. This came to be called *Fermat's Last Theorem*. For four subsequent centuries, mathematicians across the world were intrigued by Fermat's claim, trying to come up with a proof, but always to no avail, until June of 1993, when British mathematician Andrew Wiles declared that he had finally proved it. In December of that year, some mathematicians found a gap in his argument. But in October of 1994, Wiles, together with Richard L. Taylor, filled that gap to virtually everyone's satisfaction. The Wiles–Taylor proof was published in May 1995 in the *Annals of Mathematics*.

The Wiles–Taylor proof is the result of connecting and modifying previous ideas and formulas used to explore Fermat's Last Theorem. It is a brilliant proof, but Fermat's Last Theorem still haunts some mathematicians, for the simple reason that the Wiles–Taylor proof was certainly not what Fermat could have envisioned, because it depended on mathematical work subsequent to Fermat. In a sense, Fermat left behind a true mathematical mystery. What possible "simple proof" could he have been thinking of as he read Diophantus's *Arithmetica*? Was he mistaken or was his proof so different, or perhaps so simple, that it has eluded us to this day?

The Circle

Alcuin's Problem 25 deals with the circle—another figure that has been of great importance to the development of mathematics. A *circle* is defined as the set of all points in a plane that are a given distance, r, from a fixed point. Along with quadrangular and triangular figures, it was one of those geometric figures that students in antiquity and the medieval era studied in great detail, not only for their architectural and engineering applications, such as in city design, but also because they were the source of early theorems and proofs that put on display mathematical thinking at its core. The *Rhind Papyrus*, for example (Chapter 1), presents one of the first formal estimations of the value of π via a problem (number 78). It is paraphrased below:

What is the area of a circle inscribed in a square that is 9 units on its side?

It is worth breaking down the geometric principles and techniques involved in the problem for the sake of illustration. We start by drawing a diagram showing a circle inscribed in a 9 × 9 square. This means that the circumference touches the square at four points, called "tangency points." The diameter is 9 units, which can be shown by a diameter line drawn parallel to a side and touching two points of tangency (Figure 3.14).

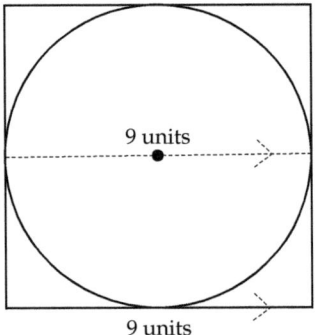

9 units

9 units

Figure 3.14 Circle in a 9 by 9 Square

Next, we trisect each side of the square—noting that each trisection is 3 units in length. This allows us to construct nine internal 3 × 3 squares as shown in Figure 3.15.

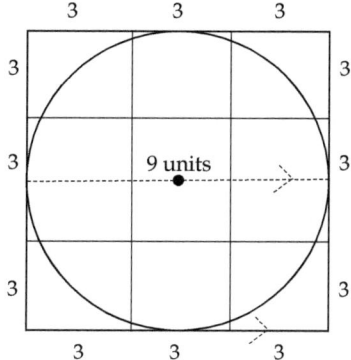

Figure 3.15 Trisection of the Square

Next, we draw diagonals in the four corner squares. This produces an octagon, as can be seen in Figure 3.16.

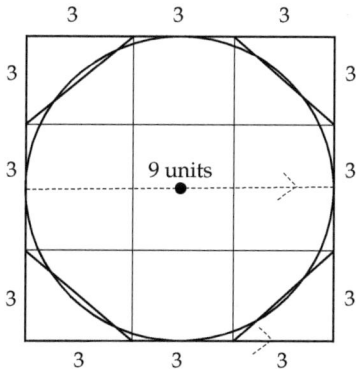

Figure 3.16 Diagonals Drawn in Corner Squares

The area of the octagon is equal to the areas of the five inner squares plus half the areas of the four corner squares. Four half squares are equal to two squares. The area of the octagon is thus equal to the sum of the areas of seven small squares (5 + 2). The area of one small square is 3 × 3 = 9 square units. The total area of seven such squares is, therefore, 9 × 7 = 63 square units. The area of the circle is a little more than this. So, let us assume that it is 64. Using contemporary notation, the estimation of π unfolds as follows:

Diameter = 9

Radius (r) = 9/2 = 4.5

r^2 = 20.25

Area of circle = πr^2 = 64

So, $\pi = 64/r^2 = 64/20.25 = 3.16049\ldots$

The circle, like the other classic geometrical figures, has been the source of many puzzles in recreational mathematics. Consider the following one, which at first glance would seem to suggest a geometry problem that can be solved routinely, but which actually requires outside-the-box thinking. It was devised by Martin Gardner (1994: 44) (Figure 3.17).

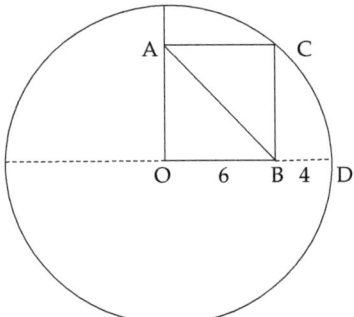

Figure 3.17 Gardner's Puzzle

Calculate the length of the diagonal AB in rectangle AOBC. Note that O is the center of the circle.

At first consideration, the solution seems to be intractable. Since the diagonals of a rectangle are equal in length, as are the radii of a circle, let us draw the other diagonal (OC) of rectangle AOBC, taking note that it is also a radius of the circle (Figure 3.18).

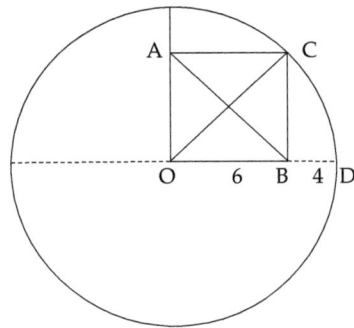

Figure 3.18 Drawing the Diagonal OC

Line OBD is a radius as well, and it is equal to 10 (= 6 + 4). So, line OC, being a radius, is also equal to 10. From this, we conclude that the other diagonal of the rectangle, AB, is 10 units in length.

Epilogue

There is little doubt that Alcuin included his ingenious geometrical problems in the *Propositiones* because he well understood that they stimulated students to think creatively, as well as showing them how geometry and numerical concepts were interrelated. In his era, geometry was seen to constitute the basis for a systematic study of mathematics, since it highlighted the methods of proof established by Euclid and other ancient thinkers. Along with arithmetic, it was considered to be a critical subject for cultivating logical and imaginative thinking in tandem.

Geometry has expanded considerably since Alcuin's times. The merger of algebra and geometry in the 1400s, for instance, produced coordinate geometry, without which there would be no advanced branches, such as the calculus and imaginary numbers. Nonetheless, these would not have come about without Euclid in the first place.

Explorations

The explorations in this section require an elementary knowledge of geometry and algebra. Several of them are classic puzzles in recreational mathematics, requiring quite a bit of imaginative thinking.

1. *A Chicken Pen Problem*

A farmer wants to fence in a chicken pen with the 60 feet of wire he has available. He uses his larger barn for one of the four sides of the pen. He wants the front side (opposite the barn side) to measure three times the length of either one of the two lateral sides. What will its dimensions be?

[Ever since Alcuin's times, this type of problem has been included in school texts across the world. Like Alcuin's problems, it is designed to get students (and others) to think geometrically.]

2. *A Pool Problem*

A contractor wanted to build a rectangular pool to measure 4 feet wide by 9 feet long, with a walk of uniform width around it. He can only afford to make the area of the walk 68 square feet. So, how wide should he make the walk?

[This type of problem based on Euclidean geometry is not unlike the type used in medieval texts such as the *Propositiones*.]

3. *A Rectangular Box Problem*

The length of a rectangular sheet of cardboard is twice its width. In each corner, a 2-inch square is drawn in outline form so that two sides of each square can be cut out, allowing for the four sides of the cardboard formed as a result of the cut-outs to be turned up to make a rectangular box. If the box has a volume of 60 cubic inches, what were the original dimensions of the cardboard?

[Note that the volume of a rectangular box is equal to its length × width × height.]

4. *Making a Rectangle*

You are given a shape which appears to be a rectangle with two tabs jutting out from it. Can you dissect it in such a way as to make an actual rectangle, without the tabs (Figure 3.19)?

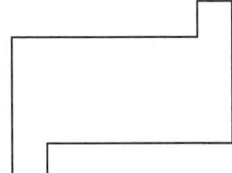

Figure 3.19 The Rectangle Puzzle

[This is an ingenious "thinking outside the box" puzzle attributed originally to magician and writer Angelo John Lewis (1839–1919), as far as can be determined. It clearly requires a large dose of imaginative geometric analysis.]

5. *Martin Gardner's Triangle Puzzle*

Given an obtuse triangle (a triangle with an angle greater than 90°), is it possible to cut the triangle into seven smaller triangles, none of which is a right triangle (a triangle in which one of its angles is 90°)?

[This is based on one of Martin Gardner's clever puzzles, see Annotations.]

6. *A Stick Puzzle*

Below is an arrangement of six sticks representing the fraction 1/7 in Roman numerals (I/VII). Can you move a single stick, other than the horizontal one, to produce a fraction equal to 1 (Figure 3.20)?

Figure 3.20 A Stick Puzzle

[Stick puzzles are a popular type of activity in recreational mathematics, involving a blend of spatial and numerical thinking. See Pieter van Delft's interesting treatment, *Creative Puzzles of the World* (1978).]

7. Disc Movement Problem

There are six discs laid out in a row on a table, three colored white and three colored black, with one space between the two color sets at the start (Figure 3.21).

Figure 3.21 Initial Layout of the Discs

Change the positions of the sets by moving only one disc at a time (Figure 3.22).

Figure 3.22 Required New Layout

You can slide a disc into an empty space, or you may move it over an adjacent disc into an empty space. You are not allowed to move a disc backward: that is, the white discs can only move rightward, and the black discs only leftward.

[Disc movement puzzles are found in many forms across continents and throughout the ages. The ancient Japanese called them *Hiroimono* ("things picked up"), because such games are played by picking up and moving things one at a time (see Costello 1988).]

8. The Bookworm Puzzle

Three book volumes, I, II, and III, are stacked upright against each other on a bookshelf. For each volume, each of its covers, front and back, is ½ of an inch thick, and the width of all the pages in between the two covers is 2 inches. A bookworm starts on the first page of Volume I and bores its way straight through to the last page of Volume III. How far has the bookworm gone?

[This is a take on a famous puzzle devised by Eugene Northrop in his classic 1944 book in recreational mathematics, titled *Riddles in Mathematics*.]

9. *Another Puzzle Based on a Martin Gardner Puzzle*

On a sheet of paper, draw 10 straight equidistant and equal parallel lines (Figure 3.23).

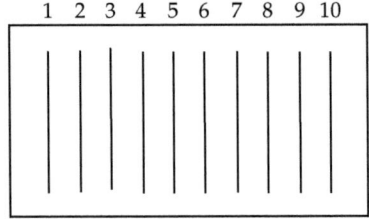

Figure 3.23 Sheet with 10 Equal and Parallel Lines

Now, cut a diagonal line through the sheet that touches the top point of line 10 and the bottom point of line 1. This produces an upper and lower part (Figure 3.24). The upper part is on the left and the lower on the right.

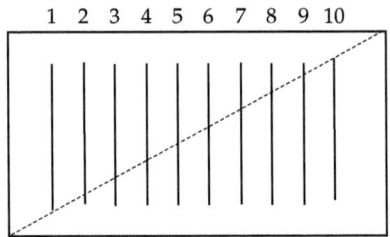

Figure 3.24 Previous Figure with Diagonal

Slide the lower part one unit to the left, producing two protruding parts, A and B (Figure 3.25).

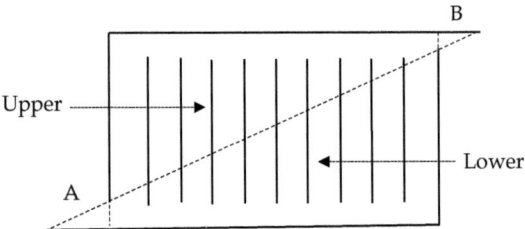

Figure 3.25 Lower Part Slide

Cut out the protruding parts and slide the upper and lower sections to realign them into a quadrangular sheet again. If we count the lines inside now, we can see that there are 9 lines, instead of 10 (Figure 3.26).

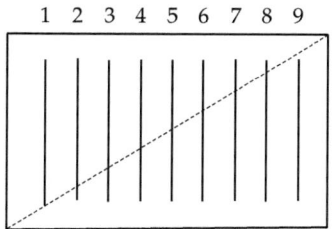

Figure 3.26 Realignment

What happened to the 10th line?

10. *The Color of the Bear*

A group of scientists, having pitched camp, set forth to go bear observation. They walk 15 miles due south, then 15 miles due east, where they see a bear. They take their photos, and then return to camp by traveling 15 miles due north. What was the color of the bear?

[This puzzle is a famous one that was apparently devised by Albert Einstein, according to some puzzle historians. It is a simple geometry puzzle, but it requires outside-the-box thinking.]

Cited Works and Further Reading

Burkholder, Peter (1993). Alcuin of York's *Propositiones ad acuendos juvenes*: Introduction, Commentary & Translation. *History of Science & Technology Bulletin*, Vol. 1, number 2.

Dantzig, Tobias (2005). *Number: The Language of Science*. New York: Plume.

Delft, Pieter van (1978). *Creative Puzzles of the World*. New York: H.N. Abrams.

Diophantus (3rd century CE). *Arithmetica*. New York: Springer, 1982.

Euler, Leonhard (1736). *Mechanica, sive motus scientia analytice exposita*. Petropoli: Typographia Academiae Scientiarum.

Euler, Leonhard (1748), *Introductio in analysin infinitorum*. Lausanne: Marcum-Michaelem Bousquet.

Frederickson, Greg N. (2003). *Dissections: Plane and Fancy*. Cambridge: Cambridge University Press.

Gardner, Martin (1956). *Mathematics Magic and Mystery*. New York: Dover.

Gardner, Martin (1983). *Wheels, Life and Other Mathematical Amusements*. New York: Freeman.

Gardner, Martin (1994). *My Best Mathematical and Logic Puzzles*. New York: Dover.

Hadley, John and Singmaster, David (1992). *Problems to Sharpen the Young*. *Mathematics Gazette* 76: 102–126.

Hooper, William (1774). *Rational Recreations*. London: Holborn.

Jones, William (1706). *Synopsis Palmariorum Matheseos, or a New Introduction to Mathematics*. London: J. Matthews.

Loyd, Sam (1914). *Cyclopedia of Tricks and Puzzles*. New York: Dover.

Netz, Reviel and Noel, William (2007). *The Archimedes Codex: Revealing the Secrets of the World's Greatest Palimpsest*. London: Weidenfeld & Nicholson.

Northrop, Eugene (1944). *Riddles in Mathematics*. London: Penguin.

O'Connor, J. J. and Robertson, E. F. (2012). *Propositiones ad acuendos juvenes*. https://mathshistory.st-andrews.ac.uk/HistTopics/Alcuin_book/.

Stewart, Ian (2008). *Taming the Infinite*. London: Quercus.

Taylor, Richard and Wiles, Andrew (1995). Ring-Theoretic Properties of Certain Hecke Algebras. *Annals of Mathematics* 141: 553–572.

4
Algebra

Prologue

The history of *algebra* began in the ancient world, although it was not named specifically in this way. One of the first to use abstract symbols for numbers was the Greek mathematician Diophantus, who lived in the third century. In his book, *Arithmetica*, he demonstrated how to solve arithmetical problems with symbols, prefiguring algebra as a way to do "generalized arithmetic" (Christianidis and Oaks 2023). As mentioned in the introduction to this book, it was Muḥammad ibn Mūsā al-Khwārizmī who actually established algebra as a distinct branch of mathematics, with his treatise *The Compendious Book on Calculation by Completion and Balancing* (c. 833).

Among Alcuin's *Propositiones* there are a series of "story problems," as we would call them today, which he likely included to get his students or readers, in general, to engage with basic algebra. There was no formal algebra as such in his days. But these problems nonetheless put on display what algebra is essentially all about—thinking about arithmetical problems in abstract ways.

The Problems

There are 11 problems in the *Propositiones* dealing with algebra—numbers 2, 3, 4, 7, 16, 36, 37, 40, 44, 45, and 48. Excluded here are Alcuin's Diophantine problems, which will be the subject matter of the next chapter. For the sake of convenience, modern-day algebraic notation will be used throughout in the discussion of the problems.

2. *Problem of the Group of Men*

A certain man was walking in the street when he saw a group of other men coming towards him, saying to them: "Suppose there were as many more of you as there are now; and then to this number half of half were added; and then again half of this number were added, together with myself, we would number 100." How many men were in the group that the man saw?

3. *Problem of the Storks*

Two men were walking along a street when they noticed some storks. They asked each other: "How many storks are there?" Discussing the situation, one of them remarked: "If the storks were doubled, and the original number added again, and then half of a third of this sum were added, then, together with another two storks, there would be 100 storks." How many storks did the men actually see?

4. *Problem of the Horses*

A man saw some horses grazing in a field noting the following: "Suppose that they were mine, and that the number were doubled, and then a half of half of this sum were added. In this case I would have 100 horses." How many horses did the man see grazing in the field?

7. *Problem of the Plate Weighing 600 Solidi*

There is a plate weighing 600 solidi. In it, there is gold, silver, brass, and tin. It has three times as much silver as gold, three times as much brass as silver, and three times as much tin as brass. How much does each type of metal weigh?

[A *solido* was a Roman coin and measure, used commonly in medieval Europe.]

16. *Problem of the Oxen*

Two men were leading oxen along a road, stopping at one point to make conversation. The first one said to the other: "If you give me two of your oxen, then I'll have as many oxen as you have." The other one then said: "Well, if you were to give me two oxen of what you would have if this were to occur, I would then have twice as many as you." How many oxen were there, and how many did each man have?

36. *Problem of an Old Man and a Boy*

An old man greeted a boy as follows: "May you live for a long time—as long as you have already lived, and then another amount equal to your age at that time, and then three times as much. And if God will grant you one more year than that, you shall live to be 100." How old was the boy at the time the old man greeted him?

37. *Problem of the Workmen*

A man wanted to build a house. So, he hired six workmen, of whom five were master builders and one was an apprentice. It was agreed between the man and the workmen that a total of $25 should be given to them per day as pay, and that the apprentice should receive half of what a master builder would receive per day. How much did each of them receive per day?

40. *Problem of the Grazing Sheep*

A man saw a flock of sheep grazing on the mountainside, saying to himself: "I wish I had that number of sheep, and then just as many more, plus a half of half of this number, and then another half of the last amount added on. If that were so, and I took the sheep back to my home together with me there would be, counting myself, 100 in total." How many sheep did the man see grazing?

44. *Problem of the Boy and His Father*

A certain boy, upon seeing his father, said to him: "Greetings, father!" His father answered: "Son, may you live to twice your present age and then three times the age you will then be. If one year is added to this, then you will live to be 100 years old." How old was the boy?

45. *Problem of the Pigeons*

A pigeon perched on a tree saw other pigeons flying by, cooing out to them: "If there were as many of you again and the same number were added again, then, along with me, the total number would be 100." How many pigeons flew by the original pigeon?

48. *Problem of the Students*

A certain man encountered a small group of students, asking them, "How many of you are there in your school?" One of the students replied: "I do not want to tell you directly, but I'll tell you how to figure it out. If you double the number of students, then triple that number, then divide that number into four parts, and, finally, add me, there will be 100." How many students are in the school?

Solutions

2. *Answer*: 36

Solution

The breakdown is as follows:

- Let the number of men in the group be represented by x.
- The statement "Suppose there were as many more of you [that is, x more] as there are now [which is x]" translates into algebraic language as $x + x$ or $2x$. This is the first "expression," as it is called in

algebra, that will make up the equation representing the total number of men.

- The statement "and then to this number [which is $2x$] half of half were added" translates as follows: $1/2 \times 1/2 \times 2x$, which is simplified to $x/2$. This is the second expression that will go into the equation.
- The statement "and then again half of this number [which is $x/2$] were added" translates as: $1/2 \times x/2 = x/4$. This is the third expression in the equation.
- The statement "together with myself" translates simply as the number "1" to be added to the equation as a fourth element in the equation, called a "numerical expression."
- The final statement "we would number 100" tells us that the four expressions in the equation equal "100":

$$2x + x/2 + x/4 + 1 = 100$$
$$2x + 3x/4 + 1 = 100$$
$$2x + 3x/4 = 100 - 1$$
$$2x + 3x/4 = 99$$
$$8x + 3x = 396$$
$$11x = 396$$
$$x = 36$$

3. *Answer*: 28

Solution

The breakdown is as follows:

- Let x be the number of storks.
- The statement "If the storks were doubled" translates algebraically as $2x$.
- The statement "and the original number [which is x] is added again [to the $2x$]" translates as $x + 2x = 3x$.
- This is the first expression in the equation: $3x$.
- The statement "half of a third of this sum [which is $3x$]" translates as $1/2 \times 1/3 \times 3x = 1/6 \times 3x = 3x/6 = x/2$.
- This is the second expression in the equation: $x/2$.
- The statement "with another two storks" translates as the simple numerical expression "2" (the third in the equation).
- Adding all these up would equal "100," as indicated by the final statement: "together. . . there would be 100 storks."

$$3x + x/2 + 2 = 100$$
$$3x + x/2 = 100 - 2$$
$$3x + x/2 = 98$$
$$6x + x = 196$$
$$7x = 196$$
$$x = 28$$

4. *Answer*: 40

Solution

Since this problem is solved in the same way as the previous two, the breakdown can be simplified as follows:

- Let x be the number of horses.
- Doubling this is represented by $2x$ (first expression).
- Half of half of $2x$ is: $1/2 \times 1/2 \times 2x = x/2$ (second expression).
- Adding these expressions together equals 100.

$$2x + x/2 = 100$$
$$4x + x = 200$$
$$5x = 200$$
$$x = 40$$

7. *Answer*: 15 solidi of gold, 45 of silver, 135 of brass, and 405 of tin

Solution

Here is the breakdown:

- Let x represent the weight of gold in solidi (first expression).
- The silver weighs three times this amount, or $3 \times x = 3x$ solidi (second expression).
- The brass weighs three times the amount of the silver, or $3 \times 3x = 9x$ solidi (third expression).
- The tin weighs three times the amount of the brass, which is $3 \times 9x = 27x$ (fourth expression).
- Altogether the metals weigh 600 solidi:

$$x + 3x + 9x + 27x = 600$$
$$40x = 600$$
$$x = 15$$

Therefore:

$x = 15$ solidi of gold
$3x = 3 \times 15 = 45$ solidi of silver
$9x = 9 \times 15 = 135$ solidi of brass
$27x = 27 \times 15 = 405$ solidi of tin

16. *Answer*: 12 oxen: one man had four oxen, the other man had eight oxen

Solution

The previous problems involved setting up one equation with expressions in one unknown (or variable), namely x. This one requires two equations in two unknowns, x and y. Note that any letter or symbol would do, not just the letters x and y.

Box 4.1 Unknowns and Variables

In mathematics, an *unknown* is, literally, a number we do not know. They are also called *variables*, represented by letters such as x, y, and z. This usage is traced to René Descartes (1596–1650) in his work, *La géométrie* (1637), where he introduced the convention of using the lowercase letters at the beginning of the alphabet for known quantities (a, b, and c) and those at the end of the alphabet for unknown quantities (x, y, and z).

Here is the breakdown:

- Let the number of oxen that the first man had originally be represented by x.
- Let the number of oxen that the second man had originally be represented by y.

First Hypothetical Transaction

- If the first man gets two oxen from the second man, he would have $(x + 2)$ oxen, while the second man would then have two less, that is, $(y - 2)$ oxen.
- The statement by the first man that he will then have "as many oxen" as the second man translates as the following equation:

$$(x + 2) = (y - 2)$$

- This is the first equation. We note that for two unknowns we need two equations. So, let us set up the second one on the basis of the statement made by the second man.

Second Hypothetical Transaction

- After the first transaction, the first man would have $(x + 2)$ oxen. If two of these are given to the second man, then he would be left with

$(x + 2) - 2 = x$ oxen. The second man, who has $(y - 2)$ oxen at this point, would add these two oxen to what he has: $(y - 2) + 2 = y$.

- This, the second man says, would be twice as many as what the first many has [which is x]. So:

$$y = 2x$$

- This is the second equation. The two equations are shown below:

$$(1)\ (x + 2) = (y - 2)$$
$$(2)\, y = 2x$$

- From (2) we can see that $y = 2x$. So, we can substitute this for y in equation (1):

$(x + 2) = (y - 2)$
$x + 2 = 2x - 2$
$2x - x = 2 + 2$
$x = 4$, which is the number of oxen that the first man has

- We can now put this value of x in equation (2):

$y = 2x$

$y = 2(4) = 8$, which is the number of oxen that the second man has

- So, together there were 12 oxen $(4 + 8 = 12)$.

Box 4.2 Simplification

The term *simplification* refers to the process of replacing a mathematical expression by an equivalent one that is simpler to work with. This applies especially to equations with fractions, such as $x/2 + 5 = 12$. Simplifying in this case is equivalent to multiplying each value by 2, which would eliminate the fraction: $x + 10 = 24$. Another common example is to combine like terms in an equation such as $5x + 2x + 3x = 94$, which becomes $10x = 94$.

---

36. *Answer*: 8 years, 3 months

Solution

Here is the breakdown:

- Let the boy's age be x.
- Living as long as he has already lived means that he should live another x years, or $x + x$ years $= 2x$.
- Another amount equal to the age of $2x$ is $2x$. So: $2x + 2x = 4x$.
- Three times as much is: $3(4x) = 12x$.
- One year more is $12x + 1$.
- This would then equal 100: $12x + 1 = 100$.
- That is the equation to be solved:

$$12x + 1 = 100$$
$$12x = 100 - 1$$
$$12x = 99$$
$$x = 8.25, \text{ which is 8 years 3 months}$$

37. *Answer*: Each master builder gets $4.54 per day and the apprentice $2.27 per day.

Solution

Here is the breakdown:

- Let x dollars represent what each master builder receives per day.
- There were five of them, so in total they received $5x$ dollars per day.
- The apprentice received 1/2 of what a master builder received per day. Since a master builder received x dollars per day, this means that the apprentice received $1/2 \times x = x/2$.
- So, the master builders and the apprentice together were paid $5x + x/2$ per day. This was equal to $25. Therefore:

$$5x + x/2 = 25$$
$$10x + x = 50$$
$$11x = 50$$
$$x = 4.54 \,(= \text{ what each master builder receives per day})$$
$$x/2 = 4.54/2 = 2.27 \,(= \text{ what the apprentice receives per day})$$

Note that the calculations do not add up exactly to $25.

Each master builder was paid $ 4.54.

So, 5 of them received $5 \times \$ 4.54 = \$ 22.70$.

The apprentice received $ 2.27.

Together: $22.70 + $2.27 = $24.97.

To resolve this, Alcuin uses a bit of cleverness, even if it is not really a solution. So, the man paid each master builder $4 per day. Since there were five of them, this means that he paid out $20. He then paid the apprentice $2 per day. Together, he paid out $20 + $2 = $22. He then paid out the remaining $3 equally to the six workmen: $3.00 ÷ 6 = $0.50 (each), which means that a master builder received $4.50 per day and the apprentice $2.50.

40. *Answer*: 36 sheep

Solution

Here is the breakdown:

- Let x be the number of sheep in the flock.
- That number (x) and "just as many more" (another x) would be:

$$x + x = 2x \text{ (first expression in the equation)}$$

- Then, "a half of half this number" ($2x$) translates as:

$$1/2 \times 1/2 \times 2x = 1/4 \times 2x = 2x/4 = x/2 \text{ (second expression in the equation)}$$

- "Another half of the last amount" ($x/2$) translates as:

$$1/2 \times x/2 = x/4 \text{ (third expression in the equation)}$$

- The man makes up the numerical expression ("1") in the equation.
- Together, the expressions add up to "100":

$$2x + x/2 + x/4 + 1 = 100$$
$$2x + x/2 + x/4 + 1 = 100 - 1$$
$$2x + x/2 + x/4 = 99$$
$$8x + 2x + x = 396$$
$$11x = 396$$
$$x = 36$$

44. *Answer*: 16 years, 6 months

Solution

Here is the breakdown:

- Let x be the boy's age in years.
- Twice his present age would be: $2x$.
- Three times that age would be: $3(2x) = 6x$.
- Adding one year to this would be: $6x + 1$.
- This would be equal to 100 years:

$$6x + 1 = 100$$
$$6x = 100 - 1$$
$$6x = 99$$
$$x = 16.5 = 16 \text{ years and } 6 \text{ months}$$

45. *Answer*: 33

Solution

Here is the breakdown:

- Let x be the number of pigeons that flew by the original pigeon.
- "As many of you again" translates as x.
- The "same number again" also translates as x.
- So, in total this would make: $x + x + x = 3x$.
- Adding the original pigeon to this amount, there would be 100 pigeons in total:

$$3x + 1 = 100$$
$$3x = 100 - 1$$
$$3x = 99$$
$$x = 33$$

48. *Answer*: 66

Solution

Here is the breakdown:

- Let x be the total number of students in the school.
- Doubling that number translates as: $2x$.
- Tripling that number ($2x$) translates as: $3(2x) = 6x$.
- Dividing that number ($6x$) by 4 translates as $6x/4 = 3x/2$.
- Adding one to this would make 100 students:

$$3x/2 + 1 = 100$$
$$3x/2 = 100 - 1$$
$$3x/2 = 99$$
$$3x = 198$$
$$x = 66$$

Annotations

Alcuin's algebra problems were probably intended to show his readers that algebra is nothing more than generalized arithmetic. Algebra was not institutionalized into mathematics at the time, as discussed. It was the Persian mathematician al-Khwārizmī, who established algebra as a distinctive method of analysis. The next major step forward was taken by two French mathematicians, François Viète (1540–1603) and René Descartes (1596–1650). Descartes' most significant achievement in this area was the use of algebraic formulas to describe geometric figures, which formed a branch of mathematics known as analytic geometry, an approach to mathematics prefigured by the eleventh-twelfth-century Persian mathematician Omar Khayyam (1048–1131).

Equations

Alcuin's problems, except one (number 16), are solved by setting up equations in one unknown. Solving such an equation means finding the value for the unknown. Equations with more than one variable, as in Problem 16, require corresponding equations to solve, called a *system* of simultaneous equations. A common method for solving them is known as the *Gauss method*, after the German mathematician Carl Friedrich Gauss. The objective is to make the numerical coefficients of one variable the same in both equations, so that the variable can be eliminated by either subtracting or adding the equations. A *coefficient* is a quantity placed before (and thus multiplying) the variable in an algebraic expression: in "$4x$," 4 is the numerical coefficient and x is the variable.

Consider the two equations below:

$$4x + 3y = 13$$
$$x + 2y = 2$$

Consider the second one. If the coefficient of x were 4, then we could subtract the equation from the first one to eliminate the x variable. So, let us multiply both sides by 4, since this does not change the value equivalency:

$$4(x + 2y) = (4)(2)$$
$$4x + 8y = 8$$

We put this equivalent equation under the first equation:

$$4x + 3y - 13$$
$$4x + 8y = 8$$

We now subtract the two equations to get:

$$3y - 8y = 13 - 8$$
$$-5y = 5$$
$$y = -1$$

Replacing y with -1 in either of the two original equations allows us to determine that $x = 4$. Take the first equation:

$$4x + 3y = 13$$
$$4x + (3)(-1) = 13$$
$$4x - 3 = 13$$
$$4x = 16$$
$$x = 4$$

Note that the value for y is a negative number, a type of number that is not found in the *Propositiones*.

Box 4.3 Negative Numbers

A negative number, a, is shown with a minus sign as "$-a$." It is a number whose value is less than zero. One of the first mentions of such a number is found in a 250 BCE Chinese text titled *Chui-chang swan-shu* ("The Nine Chapters"). During the seventh century, negative numbers were found in the bookkeeping practices and astronomical calculations of the Hindus. It was not until the sixteenth century, however, that such numbers became a regular part of arithmetic and algebra.

A few centuries after Alcuin's *Propositiones*, algebra developed its own notions, terminology, and theorems. Some of these are summarized in Table 4.1, for the sake of convenience.

The Fundamental Theorem of Algebra

By the sixteenth and seventeenth centuries, algebra had emerged as an important tool in the development of number theory and other branches of mathematics. Particularly important was the work of French mathematician Albert Girard (1595–1632), whose 1629 book *Invention nouvelle*

Table 4.1 Some notions, terminology, and symbols of algebra.

Term	Definition	Examples
Expression	A statement made up of variables and (often) constant (numerical) terms	$5x + 2, 3x^2 - 2x, 6x - y^2 \ldots$
Binomial	An expression consisting of two terms	$3x + 5, 7x^2 + 10, \ldots$
Coefficient	A number or symbol multiplying a variable quantity in an algebraic expression	The 5 in $5x^2$ or the a in ax^3; in the former it is called "numerical" and in the latter "literal"
Constant	Any number in an expression	The 7 in $(7 - x)$ or the 15 in $(15 + a^2b)$
Equation	Any equivalence relation with "two sides" separated by the "=" sign, indicating that the values of the two sides are equal	$5x + 4 = 98; 3x^3 + 2y^2 = 94; \ldots$
Monomial	An expression consisting of one term	$5x, x^5, y^3 \ldots$
Polynomial	An expression with an unspecified number of terms	$(5x^4 + 7x^2 + 4x + 98), \ldots$
Trinomial	An expression consisting of three terms	$(x + y + z), (5x^2 - x + 5), \ldots$
Variable	A letter (or symbol) standing for some number; also called the unknown	The x in $7x^5$ or the a in $5a$

en algèbre ("New Invention in Algebra") introduced the so-called *Fundamental Theorem of Algebra*. This states that every polynomial equation of degree "n" has "n" roots, or solutions. In its simplest form, this means that a linear equation will have one root, a quadratic equation (discussed in the next chapter) will have two roots, a cubic equation three, a quartic equation four, and so on:

Linear Equation

$x - 5 = 0$

One root: $x = \{ + 5 \}$

Quadratic Equation

$x^2 - 4 = 0$

Two roots: $x = \{ + 2, -2 \}$

Cubic Equation

$$x^3 - 2x^2 - 5x + 6 = 0$$

Three roots: $x = \{+1, -2, +3\}$

And so on.

However, in its complete form, the theorem states that for any polynomial equation there is at least one *complex* number root, for n solutions.

Box 4.4 Complex Numbers

Without going into details here, suffice it to say that a complex number is a number made up of real and imaginary number parts. If a and b are real numbers (such as 3, 9, 3/4, and 6/15), and i is the imaginary unit number, $\sqrt{-1}$, then the general form of a complex number, z, is: $z = a + bi$. The a is the real part and the bi the imaginary part. Any real number can thus be seen to be a particular type of complex number with $bi = 0$.

Consider the following quadratic equation, which according to the theorem should have two roots.

$$x^2 - x + 1 = 0$$

Using the quadratic formula (discussed in the next chapter), there are indeed two roots, which turn out to be:

$$x = \{(0.5 + 0.866i), (0.5 - 0.866i)\}$$

As can be seen these roots have the form of complex numbers:

$$(a + bi), (a - bi)$$

There is much more to the Fundamental Theorem, which is beyond the scope of the present discussion. The point here is that algebra has led to many new vistas in mathematics and continues to do so, even though the ideas took a long time to come to fruition. To cite mathematician Morris Kline (1967: 22): "The history of arithmetic and algebra illustrates one of the striking and curious features of the history of mathematics. Ideas that seem remarkably simple once explained were thousands of years in the making."

Epilogue

Alcuin's inclusion of algebraic problems in his *Propositiones* was intended, no doubt, to show his medieval students and other readers that the topic of equations was not the daunting subject that they might have thought it was. The main difference between carrying out an operation arithmetically and doing so algebraically lies in thinking about the operation in *general* rather than in *specific* numerical terms. In the case of story problems, algebra is a translation code, as George Polya (1957: 174) so aptly put it:

> To set up equations means to express in mathematical symbols a condition stated in words; it is translation from ordinary language into the language of mathematical formulas. The difficulties which we may have in setting up equations are difficulties of translation.

Explorations

The explorations below are based on simple algebra, in the spirit of Alcuin's own problems. There are also a few classic algebraic conundrums in the mix.

1. *Fish Size*

A farmer goes to market to buy freshly caught fish. He saw one that he really liked, thinking to himself, after measuring it: "Well it is quite a fish. It is 20 centimeters long plus half of its own length." How long was it?

2. *Flock of Sheep*

From a high hilltop, a farmer and his daughter spotted a flock of sheep down below. The farmer cried out: "Look, my daughter, there must be a hundred sheep in that flock." The daughter, who loved mathematics and was very fastidious about being numerically accurate, but who also loved to stump her family with her mathematical wizardry, answered rather matter-of-factly: "Nonsense. In order to add up to one hundred, there would have to be again as many as there are in the flock, to which half again of the original number must be added, and then to this a quarter of the original number must be added. But even after all this, there would still have to be one more sheep added on, before there would be one hundred in total." How many sheep were in the flock?

3. *A Necklace Problem*

A woman goes to a special market to buy a necklace. She sees a beautiful multi-colored one, asking the seller how many beads there were in the necklace. The seller, being somewhat mischievous answered: "It has 10 silver beads, 2 blue beads, 1 white bead, and a number of red and turquoise beads. The number of red beads equals one-third the entire number on the necklace minus three beads, and the number of turquoise beads is one-half the number of red beads plus twice the number of silver beads." How many beads were there in total?

4. *A Problem from the Rhind Papyrus*

A quantity and its 1/7th added together become 19. What is the quantity?
[The first specific use of algebraic symbolism is ascribed to Diophantus of Alexandria, discussed in the next chapter. Nonetheless, some of the problems in the *Rhind Papyrus* (see Chapter 1), written around two millennia earlier, clearly involved the use of algebraic thinking, no matter how they were solved practically. The problem above is based on Problem 24 from that work.]

5. *An Early Medieval Inheritance Problem*

A father writes in his will: "I desire my two children to receive the thousand staters of which I am possessed, but let the fifth part of the older one's share exceed by ten the fourth part of what falls to the younger one." How much will each heir receive?
[This problem comes from the sixth century (Wells 1992: 23).]

6. *A Take on Alcuin's Problem 44*

A woman says to her friend: "I am twice the age that you are now. Ten years ago your age was one third of my age then. Ten years from now, I will be twenty years older than you." How old were the two at the time of the conversation?

7. *A Work Problem*

If it takes Angela 3 hours to paint a certain room, and her friend Bernice only 2 hours to paint the same room, how long would it take to paint the room if they worked on it together (at the same time)?
[Recall that this type of problem goes back to medieval mathematics, starting with the *Greek Anthology* (Chapter 1).]

8. *An Age Problem from the Greek Anthology*

Demochares has lived a fourth of his life as a boy, a fifth as a youth, a third as a man, and 13 years in his old age. How old is he?

[This problem is paraphrased from the *Greek Anthology*.]

9. *The Two Horses Problem*

Two horses are 425 feet apart. They start trotting lazily towards each other at the same time. The first one trots constantly at 45 feet per minute, the second at 40 feet per minute. How far will each horse have traveled when they meet?

[Recreational mathematics is replete with problems such as this one.]

10. *The Price of Potatoes*

A farmer went to market to buy potatoes last week. He bought them at 92¢ per pound. If the price had been 4¢ per pound lower, he could have bought one pound more for the same money. How much did he spend for the potatoes?

Cited Works and Further Reading

Al-Khwārizmī, Muḥammad ibn Mūsā (820 CE). *Kitab al-Jabr wa-al-Muqabala* (*The Compendious Book on Calculation by Completion and Balancing*). Translation by Roshdi Rashad. London: Saqi Books, 2010.

Bombelli, Rafael (1572). *L'Algebra*. Bologna: Giovanni Rossi.

Burkholder, Peter (1993). Alcuin of York's *Propositiones ad acuendos juvenes*: Introduction, Commentary & Translation. *History of Science & Technology Bulletin*, Vol. 1, number 2.

Cardano, Gerolamo (1545). *Ars magna*. New York: Dover, 1993.

Chace, Arnold B. (1979). *The Rhind Mathematical Papyrus: Free Translation and Commentary with Selected Photographs, Transcriptions, Transliterations and Literal Translations*. Reston, VA: National Council of Teachers of Mathematics.

Christianidis, Jean and Oaks, Jeffrey (2023). *The Arithmetica of Diophantus: A Complete Translation and Commentary*. London: Routledge.

Descartes, René (1637). *La géometrie*. Paris: Presses Universitaires de France, 1996.

Girard, Albert (1629). *Invention nouvelle en algèbre*. Amsterdam: Guillaume Iansson Blaeuw.

Hadley, John and Singmaster, David (1992). Problems to Sharpen the Young. *Mathematics Gazette* 76: 102–126.

Kline, Morris (1967). *Mathematics for the Nonmathematician*. New York: Dover.

Kraitchik, Maurice (1942). *Mathematical Recreations*. New York: Dover.

Newton, Sir Isaac (1707). *Universal Arithmetick*. London: W. Johnston.

O'Connor, J. J. and Robertson, E. F. (2012). *Propositiones ad acuendos juvenes*. https://mathshistory.st-andrews.ac.uk/HistTopics/Alcuin_book/.

Polya, George (1957). *How to Solve It*. New York: Doubleday.

Wells, David (1992). *The Penguin Book of Curious and Interesting Puzzles*. Harmondsworth: Penguin.

5

Diophantine Problems

Prologue

Called the "father of algebra," Diophantus of Alexandria was born between 201 and 215 CE. In his series of books, called the *Arithmetica*, many of which are now lost, he was one of the first to deal with how to solve equations with symbols (Christianidis and Oaks 2023). It is not known if Alcuin knew about the *Arithmetica*, since he left no evidence to that effect. However, he included a set of problems in the *Propositiones* that are modeled on a type of problem found in the *Arithmetica*, now called Diophantine. A Diophantine problem involves equations in which there are more unknowns than there are equations. Diophantine problems are particularly challenging, since they entail the use of systematic trial and error, guided by precise reasoning. Diophantine analysis has had deep implications for the development of mathematics.

There is a famous Diophantine equation called the "taxicab equation" which is associated with an anecdote that was told by the mathematician G. H. Hardy (1877–1947), after going to visit the renowned mathematician Srinivasa Ramunajan (1887–1920) in 1919. It is worth repeating here (Silverman 1993):

> I remember once going to see him [Ramanujan] when he was lying ill at Putney. I had ridden in taxi-cab No. 1729, and remarked that the number seemed to be rather a dull one, and that I hoped it was not an unfavourable omen. "No," he replied, "it is a very interesting number; it is the smallest number expressible as the sum of two [positive] cubes in two different ways."

The number 1729 has since been called a "taxicab number," defined as a number expressible as the sum of two cubes:

$$a^3 + b^3 = c^3 + d^3$$

The taxicab number 1729 is the smallest such number known that satisfies this Diophantine equation:

$$1^3 + 12^3 = 9^3 + 10^3 = 1729$$

Taxicab numbers were studied by Ramanujan himself, who anticipated that they revealed hidden structures in geometry, number theory, and physics (Ono and Trebat-Leder 2016).

The Problems

There are seven Diophantine problems in the *Propositiones*—numbers 5, 32, 33, 34, 38, 39, and 47. Alcuin simply provides the answers, without explanations, presumably leaving it up to the students' imagination to figure out why a problem produced the answer that it did.

5. *Problem of the Merchant and the Pigs*

A merchant wanted to buy 100 pigs for $100. He went shopping and discovered that a boar costs $10, a sow costs $5, and two piglets can be bought for $1. How many boars, sows, and piglets can the merchant buy with $100?

32. *Problem of the Servants and Corn*

The head of a household had 20 servants. He gave them 20 measures of corn as follows: the men received three measures, the women received two measures, and the children half a measure each. How many men, women, and children servants were there in the household?

33. *Problem of Other Servants and Corn*

A head of a household had 30 servants to whom he gave 30 measures of corn as follows: the men received three measures, the women two measures, and the children half a measure each. How many men, women and children servants are there in the household?

34. *Problem of Yet Another Group of Servants and Corn*

A head of a household had 100 servants. He gave them 100 measures of corn as follows: the men received three measures, the women two, and the children half a measure each. How many men, women, and children servants were there in the household?

[Note that there is more than one answer to this problem.]

38. *Problem of the 100 Animals*

A certain man bought 100 animals for $100. He paid $3 per horse, $1 per cow, and $1 per 24 sheep. How many horses, cows, and sheep did he buy?

39. *Problem of the Traveling Merchant*

A certain merchant bought 100 assorted animals for $100 on a trip to a faraway land. He paid $5 for each camel, $1 for each donkey, and $1 for 20 sheep. How many camels, donkeys, and sheep did the merchant buy?

47. *Problem of the Clergy and the Loaves*

A certain bishop ordered 12 loaves of bread to be divided amongst the clergy. He stipulated that each priest should receive two loaves, each deacon half a loaf, and each reader a quarter of a loaf. It turned out that the number of clerics and the number of loaves were the same. How many priests, deacons, and readers must there have been?

Solutions

5. *Answer*: 1 boar, 9 sows, and 90 piglets

Solution

Here is the breakdown:

- Let x be the number of boars, y the number of sows, and z the number of piglets. There were 100 pigs in total. So, the number of boars plus the number of sows, plus the number of piglets is as follows:

$$x + y + z = 100$$

- Since a boar costs $10, the total cost of the boars is $10x$.
- A sow costs $5, so the total cost of the sows is $5y$.
- Since two piglets cost $1, each piglet costs 1/2 dollar, or 0.50 cents. So, the total cost of the piglets is $0.50z$.
- The total price to be spent on the pigs is $100, which is made up of the separate costs of the boars, sows, and piglets:

$$10x + 5y + .50z = 100$$

- With the first equation, we can express z in terms of the other two variables:

$$x + y + z = 100$$
$$z = 100 - (x + y)$$
$$z = (100 - x - y)$$

- We plug this into the second equation:

$$10x + 5y + 0.50z = 100$$
$$10x + 5y + 0.50(100 - x - y) = 100$$
$$10x + 5y + 50 - 0.50x - 0.50y = 100$$
$$(10x - 0.50x) + (5y - 0.50y) + 50 = 100$$
$$(10x - 0.50x) + (5y - 0.50y) = 100 - 50$$
$$(10x - 0.50x) + (5y - 0.50y) = 50$$
$$9.5x + 4.5y = 50$$
$$95x + 45y = 500$$
$$19x + 9y = 100$$
$$19x = 100 - 9y$$
$$x = (100 - 9y)/19$$

- Now, the expression $(100–9y)/19$ has to be an integral (whole number) value, since a fractional number of pigs would make no sense.
- So, let us try a few values for y in the expression $(100 - 9y)/19$ to see which one produces an integral value:

y	(100 – 9y)/19	Integral Value?
1	4.78	no
2	4.31	no
3	3.84	no
4	3.36	no
5	2.89	no
6	2.42	no
7	1.94	no
8	1.47	no
9	1	yes

- As can be seen, the value of y for which we get an integral value is $y = 9$. This tells us the number of sows.
- We can now put this value in the equation above to solve for x:

$$x = (100 - 9y)/19$$
$$19x = 100 - (9)(9)$$
$$19x = 100 - 81$$
$$19x = 19$$
$$x = 1$$

- This tells us the number of boars. We can now plug in the values of x and y in the first equation above to solve for z, to find out the number of piglets:

$$x + y + z = 100$$
$$1 + 9 + z = 100$$
$$z + 10 = 100$$
$$z = 100 - 10$$
$$z = 90$$

32. *Answer*: 1 man, 5 women, and 14 children

Solution

Here is the breakdown.

- Let x be the number of men, y the number of women, and z the number of children.
- There were 20 servants in total, so the number of men, women, and children add up to this number:

$$x + y + z = 20$$

- Each man received 3 measures of corn, and so the total for the men was $3x$.
- Each woman received 2 measures of corn, and so the total for the women was $2y$.
- Each child received 1/2 measure of corn, and so the total for the children was $1/2 \times z = z/2$.
- There were 20 measures of corn, so the number of measures handed out to the men, women, and children added up to this number:

$$3x + 2y + z/2 = 20$$
$$6x + 4y + z = 40$$

- With the first equation, we can express z in terms of the other two variables:

$$x + y + z = 20$$
$$z = 20 - x - y$$

- We can now put this into the second equation above:

$$6x + 4y + z = 40$$
$$6x + 4y + 20 - x - y = 40$$

$$5x + 3y + 20 = 40$$
$$5x + 3y = 40 - 20$$
$$5x + 3y = 20$$
$$5x = (20 - 3y)$$
$$x = (20 - 3y)/5$$

- The expression $(20-3y)/5$ has to be an integral value, since a fractional number of servants would make no sense.
- So, as in the previous problem, let us try a few values for y to see which one produces an integral value for the expression:

y	$(20-3y)/5$	*Integral Value?*
1	3.4	no
2	2.8	no
3	2.2	no
4	1.6	no
5	1	yes

- As can be seen the value of y for which we get an integral value is $y = 5$. This tells us the number of women servants.
- We can now determine the value for x, which is the number of men servants, using the above equation:

$$x = (20-3y)/5$$
$$5x = 20-3y$$
$$5x = 20-3(5)$$
$$5x = 20-15$$
$$5x = 5$$
$$x = 1$$

- We can now plug in the values for x and y in the first equation to determine the value of z, the number of children servants.

$$x+y+z = 20$$
$$1+5+z = 20$$
$$z+6 = 20$$
$$z = 20-6$$
$$z = 14$$

33. *Answer*: 3 men, 5 women, and 22 children

Solution

Clearly, Alcuin seems to prefer this type of problem. Perhaps he may have thought that, through repetition, skill at solving Diophantine problems gradually becomes easier. Needless to say, the reasoning is the same as for the previous ones.

- Let the number of men be x, women y, and children z.

- All told, there are 30 servants:

$$x + y + z = 30$$

- The men received 3 measures each, which is $3x$ in total.
- The women 2 measures each, which is $2y$ in total.
- And the children received 1/2 measure each, for a total of $z/2$.
- Adding these up the total comes to 30 measures:

$$3x + 2y + z/2 = 30$$
$$6x + 4y + z = 60$$

- With the first equation, we can express z in terms of the other two variables:

$$x + y + z = 30$$
$$z = 30 - x - y$$

- We can now put this into the second equation:

$$6x + 4y + z = 60$$
$$6x + 4y + 30 - x - y = 60$$
$$5x + 3y + 30 = 60$$
$$5x + 3y = 60 - 30$$
$$5x + 3y = 30$$
$$5x = (30 - 3y)$$
$$x = (30 - 3y)/5$$

- We now try out a few values for y in $(30 - 3y)/5$ to see which one produces an integral value, as in the previous problems:

y	$(30 - 3y)/5$	*Integral Value?*
1	5.4	no
2	4.8	no
3	4.2	no
4	3.6	no
5	3	yes

- The first value of y for which we get an integral value is $y = 5$. This tells us the number of women servants.

- To find out the number of men servants, x, we can use the equation above, substituting 5 for y:

$$x = (30 - 3y)/5$$
$$5x = 30 - 3y$$
$$5x = 30 - 3(5)$$
$$5x = 30 - 15$$
$$5x = 15$$
$$x = 3$$

- To find out the number of children servants, z, we plug in the numerical values for x and y into the first equation:

$$x + y + z = 30$$
$$3 + 5 + z = 30$$
$$z + 8 = 30$$
$$z = 30 - 8$$
$$z = 22$$

34. *Answer*: 17 men, 5 women, and 78 children; *or* 14 men, 10 women, and 76 children; *or* 11 men, 15 women, and 74 children; *or* 8 men, 20 women, and 72 children; *or* 5 men, 25 women, and 70 children; *or* 2 men, 30 women, and 68 children

Solution

This is the exact same type of problem as the previous ones, with a difference—it produces various positive integral solutions.

- Let the number of men be x, the women y, and the children z. The total number is 100, so:

$$x + y + z = 100$$

- The men received 3 measures each, which is $3x$ in total.
- The women received 2 measures each, which is $2x$ in total.
- And the children received 1/2 measure each, which is $z/2$ in total.
- All told, they received 100 measures. So:

$$3x + 2y + z/2 = 100$$
$$6x + 4y + z = 200$$

- Consider the first equation:

$$x + y + z - 100$$
$$z = 100 - x - y$$

- We can put this into the second equation:

$$6x + 4y + z = 200$$
$$6x + 4y + 100 - x - y = 200$$
$$5x + 3y + 100 = 200$$
$$5x + 3y = 200 - 100$$
$$5x + 3y = 100$$
$$5x = (100 - 3y)$$
$$x = (100 - 3y)/5$$

- We then try out values for y in $(100 - 3y)/5$ to see which one produces an integral value, and we will find, this time, that there are six positive integral solutions:

Solution (1)

$$\text{If } y = 5$$
$$x = (100 - 3y)/5$$
$$x = (100 - 15)/5$$
$$x = 85/5$$
$$x = 17$$

Using the same substitution method of the previous problems:

$$z = 78$$

Solution set (1): $\{x, y, z\} = \{17, 5, 78\}$

Solution (2)

$$\text{If } y = 10$$
$$x = (100 - 3y)/5$$
$$x = (100 - 30)/5$$
$$x = 70/5$$
$$x = 14$$

Using the same substitution methods of the previous problems:

$$z = 76$$

Solution set (2): $\{x, y, z\} = \{14, 10, 76\}$

Solution (3)

$$\text{If } y = 15$$
$$x = (100 - 3y)/5$$
$$x = (100 - 45)/5$$
$$x = 55/5$$
$$x = 11$$

Using the same substitution method of the previous problems:

$$z = 74$$

Solution set (3): $\{x, y, z\} = \{11, 15, 74\}$

Solution (4)

> If $y = 20$
> $x = (100 - 3y)/5$
> $x = (100 - 60)/5$
> $x = 40/5$
> $x = 8$

Using the same substitution method of the previous problems:

$$z = 72$$

Solution set (4): $\{x, y, z\} = \{8, 20, 72\}$

Solution (5)

> If $y = 25$
> $x = (100 - 3y)/5$
> $x = (100 - 75)/5$
> $x = 25/5$
> $x = 5$

Using the same substitution method of the previous problems:

$$z = 70$$

Solution set (5): $\{x, y, z\} = \{5, 25, 70\}$

Solution (6)

> If $y = 30$
> $x = (100 - 3y)/5$
> $x = (100 - 90)/5$
> $x = 10/5$
> $x = 2$

Using the same substitution method of the previous problems:

$$z = 68$$

Solution set 6: $\{x, y, z\} = \{2, 30, 68\}$

38. *Answer*: 23 horses, 29 cows, and 48 sheep

Solution

The method for solving this problem is, again, the same as it is for the previous ones.

- Let x be the number of horses, y the number of cows, and z the number of sheep.
- The total number is:

$$x + y + z = 100$$

- The cost of a horse was $3, so the total cost of the horses was $3x$.
- The cost of a cow was $1, so the total cost of the cows was y.
- The cost of the sheep was $1 per 24 sheep, or $1/24 \times z$, so the total cost of the sheep was $z/24$.
- The total cost of the three animals together was:

$$3x + y + z/24 = 100$$
$$72x + 24y + z = 2400$$

- From the first equation:

$$x + y + z = 100$$
$$z = 100 - x - y$$

- We put this into the second equation:

$$72x + 24y + z = 2400$$
$$72x + 24y + 100 - x - y = 2400$$
$$71x + 23y = 2300$$
$$x = (2300 - 23y)/71$$

- Trying out various possibilities for y which produce an integral value for the expression $(2300 - 23y)/71$, it will be found that this is produced when $y = 29$:

$$\text{If } y = 29$$
$$x = (2300 - 23y)/71$$
$$x = (2300 - 667)/71$$
$$x = 23$$

Using the same substitution method of the previous problems:

$$z = 48$$

Solution set: $\{x, y, z\} = \{23, 29, 48\}$

39. *Answer*: 19 camels, 1 donkey, and 80 sheep

Solution

This is more of the same.

- Let x stand for the number camels, y for the number of donkeys, and z the number of sheep.
- The total number is:

$$x + y + z = 100$$

- The cost of a camel was \$5, so the total cost of the camels was $5x$.
- The cost of a donkey was \$1, so the total cost of the donkeys was y.
- The cost of a sheep was \$1 per 20 sheep, or $1/20 \times z$, so the total cost of the sheep was $z/20$.
- The total cost for the three animals was:

$$5x + y + z/20 = 100$$
$$100x + 20y + z = 2000$$

- From the first equation:

$$x + y + z = 100$$
$$z = 100 - x - y$$

- We put this into the second equation:

$$100x + 20y + z = 2000$$
$$100x + 20y + 100 - x - y = 2000$$
$$99x + 19y = 1900$$
$$x = (1900 - 19y)/99$$

The only integral solution set emerges when $y = 1$:

$$\text{If } y = 1$$
$$x = (1900 - 19y)/99$$
$$x = (1900 - 19)/99$$
$$x = 1881/99$$
$$x = 19$$

Using the same substitution method of the previous problems:

$$z = 80$$

Solution set: $\{x, y, z\} = \{19, 1, 80\}$

47. *Answer*: five priests, one deacon, and six readers

Solution

Here is the breakdown.

- Let x be the number of priests, y the number of deacons, and z the number of readers.
- The total number of clergy is 12, because Alcuin tells us that the number of clergy and the number of loaves is the same, namely, 12. So:

$$x + y + z = 12$$

- Each priest receives two loaves; so x priests receive $2x$ loaves in total.
- Each deacon receives 1/2 loaf; so y deacons receive $y/2$ loaves.
- Each reader receives 1/4 loaf; so z deacons receive $z/4$ loaves.
- The total number of loaves was:

$$2x + y/2 + z/4 = 12$$
$$8x + 2y + z = 48$$

- From the first equation:

$$x + y + z = 12$$
$$z = 12 - x - y$$

- We put this in the second equation:

$$8x + 2y + z = 48$$
$$8x + 2y + 12 - x - y = 48$$
$$7x + y = 36$$
$$x = (36 - y)/7$$

- The only integral solution set emerges when $y = 1$:

$$\text{If } y = 1$$
$$x = (36 - y)/7$$
$$x = 35/7$$
$$x = 5$$

Using the same substitution method of the previous problems:

$$z = 6$$

Solution set: $\{x, y, z\} = \{5, 1, 6\}$

Annotations

As with many mathematicians of antiquity, very little is known about the life of Diophantus, who lived in Alexandria, Egypt, between 200 and 298 CE. His *Arithmetica* is the work that prefigured the development of algebra and, as such, is one of the most prominent works of Greek mathematics. It is essentially, a collection of problems, based on equations, for which he provides novel ways of solving them, including the use of symbols for unknowns. While the *Rhind Papyrus* contains suggestions of symbol-based notation, and while there is some evidence that the use of symbols for numbers occurred in ancient mathematics across the world (Cajori 1909), the first systematic use of such symbolism is traced to the *Arithmetica*. As historian Kurt Vogel (1970) has noted: "The symbolism that Diophantus introduced for the first time, and undoubtedly devised himself, provided a short and readily comprehensible means of expressing an equation. Since an abbreviation is also employed for the word 'equals,' Diophantus took a fundamental step from verbal algebra towards symbolic algebra."

It is not known if Alcuin had read the *Arithmetica*, but the fact that he included Diophantine problems in his *Propositiones* suggests that, at the very least, he was familiar with the kinds of concepts and methods it entailed.

Quadratic Equations

Diophantus also examined *quadratic equations*. A linear equation is one in which the power of the variable is "1": $x^1 + 4 = 6 \rightarrow x + 4 = 6$. A *quadratic equation* is an equation, in which the power of the variable is "2": $x^2 + 4 = 20$. A quadratic equation can have second- and first-degree variables in it, but not variables to a higher power. The standard form of the quadratic equation is as follows:

$$ax^2 + bx + c = 0$$

There is evidence that quadratic equations were known in ancient Babylonia, Egypt, Greece, China, and India. The ancient mathematicians normally solved them with geometric methods. The Hindu astronomer Aryabhata (c. 476–550 CE) provided formal methods for solving quadratic equations in his work called the *Aryabhatiya* (499 CE). A little later, in the seventh century, other Indian mathematicians, notably Brahmagupta (598–668 CE), devised ingenious ways to solve quadratic equations. A few centuries after, al-Khwārizmī's treatise of algebra contained the first in-depth treatment of these equations—a treatment that was elaborated in

1545 by Gerolamo Cardano and then Simon Stevin in 1594. In 1637, René Descartes put forth the general quadratic formula that we use today in *La géométrie*:

$$x = \frac{-b \pm \sqrt{b^2 - 4ac}}{2a}$$

Let us apply this formula to the solution of the equation $x^2 + 8x + 15 = 0$, in which $a = 1$, $b = 8$, and $c = 15$:

$$\frac{-8 \pm \sqrt{8^2 - (4)(1)(15)}}{(2)(1)}$$

$$\downarrow$$

$$\frac{-8 \pm \sqrt{64 - 60}}{2}$$

$$\downarrow$$

$$\frac{-8 \pm \sqrt{4}}{2}$$

$$\downarrow$$

$$\frac{-8 \pm 2}{2}$$

$$\downarrow$$

$$(-8 + 2)/2 = -6/2 = -3$$

Or:

$$(-8 - 2)/2 = -10/2 = -5$$

Solution set:

$$x = \{-3, -5\}$$

Another way to solve quadratic equations is with the method of factoring. This means reconstructing the factors of an equation such as the one above.

$$x^2 + 8x + 15 = 0$$
$$(x + 5)(x + 3) = 0$$

So, either factor is equal to 0 :

$$(x + 5) = 0$$
$$x = -5$$
$$(x + 3) = 0$$
$$x = -3$$
$$x = \{-3, -5\}$$

Not all equations can be factored. In such cases, the general formula can be used.

Diophantus's Age Problem

One of the most famous early algebraic problems in recreational mathematics is about Diophantus's age. Ironically, it is neither a Diophantine problem, nor was it devised by Diophantus. But it is worth revisiting here, given that it is referred to as Diophantus's epitaph. It is found in the *Greek Anthology* (Introduction). It is paraphrased below:

Diophantus' boyhood lasted 1/6 of his life; his beard grew after 1/12 more; he married after 1/7 more; and his son was born 5 years later; the son lived to half his father's age, and the father died 4 years after the son. How old was Diophantus when he died?

The breakdown is as follows:

- Let Diophantus' age be x.
- His boyhood lasts 1/6 of his life = $x/6$.
- His beard grew after 1/12 more = $x/12$.
- He married after 1/7 more = $x/7$.
- At this point, Diophantus had lived $x/6 + x/12 + x/7 = 14x/84 + 7x/84 + 12x/84 = 33x/84$ years.

- His son was born 5 years later, making Diophantus $33x/84 + 5$ years old.
- The son lived to half of Diophantus's age, or $x/2$, which can be added on: $33x/84 + 5 + x/2 = 33x/84 + 42x/84 + 5 = 75x/84 + 5$.
- Diophantus died 4 years after, which made his age, $75x/84 + 5 + 4 = 75x/84 + 9$.
- So, the length of his life, x, equaled this: $x = 75x/84 + 9$.
- Solving for x we get 84, which is presumed to be Diophantus's age at his death:

$$x = 75x/84 + 9$$
$$84x = 75x + 756$$
$$9x = 756$$
$$x = 84$$

Epilogue

Whether Alcuin had read the *Arithmetica* or not, it is clear that he understood the essential nature of Diophantine thinking, which he seemingly wanted to impart to his readers as a useful tool in the conduct of everyday affairs that required some mathematical understanding. Interestingly, in a manual published in China in the fifth century, written by Zhang Quijian, one finds an identical type of problem. It reads as follows (Lam 1997):

> If a rooster is worth 5 coins, a hen 3 coins, and three chickens together 1 coin, how many roosters, hens and chickens, 100 in total, can be bought for 100 coins?

Using the same reasoning as for Alcuin's problems, but without going into details here, three solution sets are possible, with the numbers standing respectively for roosters, hens, and chickens: {4, 18, 78}, {8, 11, 81}, and {12, 4, 84}. There is no reason to believe that Alcuin was aware of Zhang Quijian's text, given that there were virtually no communications at the time between China and Europe. Moreover, it is unlikely that he could read Chinese if, somehow, he had come across the text, unless it had been translated into Latin (which is also unlikely). Actually, since identical versions of the same type of problem appear in other texts, from Diophantus onward, the likely conclusion to be drawn is that the problem reveals an archetypal conceptualization of the same kind of situation, expressed across cultures and across time in strikingly similar mathematical ways.

Explorations

The explorations require the use of either Diophantine equations or non-Diophantine quadratic equations. There are some classic puzzles interspersed into the set.

1. *The Price of Buttons*

Yesterday, a farmer's daughter went to market to buy some buttons for a sweater she was knitting. Being mathematically inclined, she thought to herself, after making the purchase: "It looks like I bought two less buttons than the price of each button." If she spent $4.83 altogether, how much did each button cost?

2. *Grandmother's Age*

A grandmother in a small medieval village has a large family. Being mathematically inclined herself, she made the following statement to another villager: "My age is between 50 and 70. Each of my sons has as many sons as brothers. The combined number of my sons and grandsons equals my age." Can you figure out her age?

3. *The Sailors, Coconuts, and the Monkey Puzzle*

Three sailors are stranded on an island where they find a pile of coconuts and a monkey. They agree to sleep that night and then divide the coconuts among themselves in the morning. During the night, one sailor wakes up, gives one coconut to the monkey, takes 1/3 of the remaining coconuts, and falls back asleep. Then a second sailor wakes up and does the same. A little later, the third sailor wakes up and does the same. When the three sailors wake up next morning, they find that there are fewer than 10 coconuts left, which they divide equally amongst themselves. How many coconuts were in the original pile?

[This is a famous puzzle based on a story published in the *Saturday Evening Post* of 1926.]

4. *One of Diophantus's Problems*

What number must be added to 100 and to 20 (the same number to each) so that the sums are in the ratio 3:1?

[This is a paraphrase of one of Diophantus's own problems.]

5. *A Digit Problem Based Again on Diophantus*

The sum of the squares of two positive consecutive integers is equal to the square of the next positive integer. What are the three integers?

[Recall Alcuin's problem about the impossibility of adding odd numbers to produce an even number (Chapter 2, Problem 43). This problem is a version, but this one has a solution. It is based directly on a famous problem in the *Arithmetica*.]

6. *Another Take on the Same Type of Problem*

In a set of three positive non-consecutive integers, the square of the largest integer is equal to the sum of the squares of the other two. The largest integer is greater by 9 than the smallest one and the other integer is greater by 7 than the smaller one. What are the three integers?

[This problem is a version of the previous problem; as something that Diophantus might have created.]

7. *The Farmer's Field*

A farmer has 768 small bushes that he wants to plant in his huge field, dividing the bushes into rows, so that the number of bushes in each row is eight more than the number of rows. What arrangement has the farmer in mind?

8. *Algebraically Guessed Age*

A merchant on his way to buy sheep at a market runs into another merchant, whom he had met the day before. Being mathematically inclined the first merchant said to the second one: "I can easily tell your age, without you telling me." "How so?" asked the second merchant. "I can do so with a little algebra. First add five to your age and then multiply the result you get by your age. Tell me what number you get from the calculations and I will then reveal your age." The second merchant carried out the required calculations and then said: "The number I came up with is 1050." "Well," said the first merchant, "I can now easily figure out your age." What was the second merchant's age?

9. *Newton's Problem*

A merchant has a certain sum of money. During the first year, he spent 100 pounds. To the remaining sum, he then added one-third of it. During the next year, he again spent 100 pounds. And increased the remaining sum by one-third of it. During the third year, he again spent 100 pounds. After he added to the remainder one-third of it, his capital was twice the original amount. What was the original sum?

[This problem was devised by Sir Isaac Newton in his *Arithmetica Universalis* (1707). The version here simply updates the language used.]

10. *One Final Diophantine Conundrum*

Multiplying the two successive integers of a number produces eight times that number plus 2. What number is it?

[Again, this type of problem is something that might have been created by Diophantus, who was among the first to develop methods for solving quadratic equations.]

Cited Works and Further Reading

Al-Khwārizmī, Muḥammad ibn Mūsā (820 CE). *Kitab al-Jabr wa-al-Muqabala* (*The Compendious Book on Calculation by Completion and Balancing*). Translation by Roshdi Rashad. London: Saqi Books, 2010.

Bachet, Claude Gaspard (1621). *Diophanti Alexandrini Arithmeticorum*. Toulouse: Bernardus Bosc.

Bashmakova, Isabella G. (1997). *Diophantus and Diophantine Equations*. Washington, D.C.: Mathematical Association of America.

Burkholder, Peter (1993). Alcuin of York's *Propositiones ad acuendos juvenes*: Introduction, Commentary & Translation. *History of Science & Technology Bulletin*, Vol. 1, number 2.

Cajori, Florian (1909). *A History of Mathematics*. London: Macmillan.

Cardano, Gerolamo (1545). *Ars magna*. New York: Dover, 1993.

Christianidis, Jean and Oaks, Jeffrey (2023). *The Arithmetica of Diophantus: A Complete Translation and Commentary*. London: Routledge.

Descartes, René (1637). *La géometrie*. Paris: Presses Universitaires de France, 1996.

Hadley, John and Singmaster, David (1992). Problems to Sharpen the Young. *Mathematics Gazette* 76: 102–126.

Lam, Lay Yong (1997). Zhang Quijian suanjing (The Mathematica Classic of Zhang Quijian): An Overview. *Archive for History of Exact Sciences* 50: 201–240.

Newton, Sir Isaac (1707). *Arithmetica Universalis*. Cambridge: University of Cambridge.

O'Connor, J. J. and Robertson, E. F. (2012). *Propositiones ad acuendos juvenes*. https://mathshistory.st-andrews.ac.uk/HistTopics/Alcuin_book/.

Ono, Ken and Trebat-Leder, Sarah (2016). The 1729 K3 Surface. *Research in Number Theory* 2, https://link.springer.com/article/10.1007/s40993-016-0058-2.

Robert of Chester (1145). *Algebra of al-Khowarizmi*. London: Macmillan, 1915.

Silverman, Joseph H. (1993). Taxicabs and Sums of Two Cubes. *American Mathematical Monthly* 100: 331–340.

Singmaster, David (1999). Some Diophantine Recreations. In: E. Berlekamp and Rodgers, T. (eds.), *The Mathemagician and Pied Puzzler: A Collection in Tribute to Martin Gardner*, 219–235. Natick, Mass.: A. K. Peters.

Stevin, Simon (1594). *De Stercktenbouwing*. Leiden.

Vogel, Kurt (1970). Diophantus of Alexandria. In: *Dictionary of Scientific Biography*. New York: Scribner.

6

Recreational Logic

Prologue

Consider the following puzzle.

> A man hiding behind a wall says to a young boy on the other side: "Do you know who I am? Your mother's mother is my mother-in-law." Who is the man in relation to the boy?

The man is the boy's father. Figuring this out involves imagining family relations in a logical hierarchical configuration. The diagram below (Figure 6.1) is one possible representation of the logic involved. In it, note that the grandmother's daughter is the boy's mother and his father's wife, and that the grandmother is his father's mother-in-law.

Now, what does this kind of problem have to do with mathematics? Alcuin certainly believed that it had a lot to do with it, since he included an identical type of problem in his *Propositiones*, Problem 11, perhaps suggesting to readers that such logical thinking is crucial to much of mathematics (Codd 1969). There is also another problem, Problem 14, that plays mischievously on logic, duping the reader into reaching a seemingly logical solution that is actually impossible.

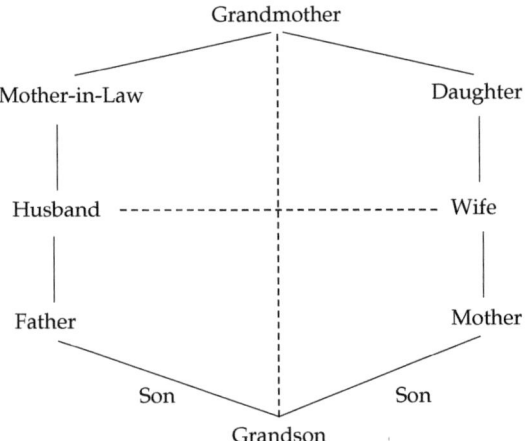

Figure 6.1 Family Hierarchy

Devoting an entire chapter, based on two problems, may seem to be unwarranted. But the type of logical thinking that Alcuin's two problems entail is fundamentally mathematical, as will be discussed. Moreover, the problems are original to Alcuin, since no earlier versions are known. They have become staples of recreational logic.

The Problems

The two problems in recreational logic are number 11 and number 14.

11. *Problem of Two Men Marrying Each Other's Sister*

If two men should marry one another's sister, what will be the sons' relations to each other?

14. *Problem of the Ox*

How many footprints are left by an ox in the last furrow after plowing all day?

Solutions

11. *Answer*: cousins
Solution
The sons are cousins twice over, because each has a parent who is a sibling to a parent of the other, in two ways. If we represent the men as M_1 and M_2 and the respective sisters of each one as S_1 and S_2, then the marriage produces the following couples:

$$M_1 - S_2$$
$$M_2 - S_1$$

The sons to whom they give birth can be represented as s_a and s_b:

$$M_1 - S_2 \rightarrow s_a$$
$$M_2 - S_1 \rightarrow s_b$$

As can be seen, the two sons are cousins twice. First, S_2 is an aunt to s_b and S_1 is an aunt to s_a, which makes the two sons cousins through this kinship line:

$$M_1 - S_2 \rightarrow s_a$$
$$\diagdown\!\!\!\!\diagup$$
$$M_2 - S_1 \rightarrow s_b$$

Second, M_1 is an uncle to s_b and M_2 is an uncle to s_a, which makes the two sons cousins again:

$$M_1 - S_2 \rightarrow s_a$$
$$M_2 - S_1 \rightarrow s_b$$

14. *Answer*: None
Solution
This is a trick question. There are no footprints in the furrow since the plow pulled by the ox behind will obliterate them.

Annotations

Solving Alcuin's kinship problem with the aid of a diagram is a long-established technique in logical analysis. One could have solved the problem without the diagram, of course. But the way in which the diagram displays relationships brings out the main reason why diagrams are so useful in logic and mathematics. Logic was a prominent topic in Greek philosophy and mathematics. René Descartes (1637) was so taken by the power of logical thinking that he even refused to accept any belief, even the belief in his own existence, unless he could "prove" it to be logically true. Descartes also maintained that logic was the only effective way to solve all human problems, most of which were caused by the emotions and the passions. Logic, for Descartes, had the ability to tame them. In their insightful book, *Descartes' Dream*, Davis and Hersh (1986: 7) encapsulated Descartes' vision as "the dream of a universal method whereby all human problems, whether of science, law, or politics, could be worked out rationally, systematically, by logical computation." German mathematician Gottfried Wilhelm Leibniz saw logic as a language of thought, recalling the origin of *logic* in the Greek word *lógos*, meaning both "word" and "thought." Leibniz called this language a *characteristica universalis*, a "universal character (symbolic) language," that could be used to great advantage for the betterment of the human condition because "errors" in thinking could be reduced to errors in the symbols and thus easily fixed (Leibniz 1667).

But what is logic? Is the logic that applies to proving theorems in mathematics the same logic that we use to solve everyday practical problems? The philosopher Charles Peirce (in his writings, 1931–1958) differentiated between two kinds of logic—*logica utens* (a practical instinctive logic) and *logica docens* (a theoretical or learned logic). The former is a rudimentary *logic-in-use* that everyone possesses without being able to specify what it is; the latter is a sophisticated and tutored use of logic practiced by mathematicians, scientists, detectives, and medical doctors. Because everyone possesses *logica utens*, no special training is required to understand what logic puzzles—such as Alcuin's kinship problem—are about or what to do in order to solve them. However, understanding formal logical structures or theories requires *logica docens*. The transformation of one into the other, the practical into the theoretical, is at the core of mathematical thinking.

The Liar Paradox

In antiquity, paradoxes emerged in philosophical debates to challenge the foundations on which the edifice of logic and mathematics was being built. One of these was the so-called *Liar Paradox*. It is the version attributed to the poet Epimenides (sixth century BCE) that has become the most widely cited one, revolving around the fact that he was a Cretan:

All Cretans are liars. Do I speak the truth?

Let us assume that Epimenides spoke the truth. Thus, his statement that "all Cretans are liars" is a true statement. However, from this we must deduce that Epimenides, being a Cretan, is a liar. But this is a contradiction. How can a liar speak the truth? Obviously, we must discard our assumption. Let us assume the opposite, namely that Epimenides is, in fact, a liar. But, then, if he is a liar, the statement he made—"All Cretans are liars"— is true. This is again a contradiction—liars do not make true statements. Obviously, we are confronted with a logical circularity.

As the famous British puzzle-maker, Henry E. Dudeney, perceptively observed in his book, *The Canterbury Puzzles* (1919), paradoxes of this kind are interesting in themselves, shedding light on the penchant for absurd or contradictory ideas, so as to see where they lead logically (if anywhere). He gave his own example:

A child asked, "Can God do everything?" On receiving an affirmative reply, she at once said: "Then can He make a stone so heavy that He can't lift it?"

The child's question is similar to a classic philosophical conundrum: What would happen if an irresistible moving body came into contact with an immovable body? As Dudeney went on to observe, such paradoxes arise only because we take delight in inventing them. In actual fact, as he put it, "if there existed such a thing as an immovable body, there could not at the same time exist a moving body that nothing could resist."

Box 6.1 Henry Ernest Dudeney (1857–1930)

Henry Dudeney was a famous, ingenious British puzzle-maker. At age nine, he started inventing difficult puzzles that he published in a local paper, under the pseudonym of "Sphinx." In 1893, he started a correspondence with his American counterpart, Sam Loyd—the two leading puzzle-makers of the day. But Dudeney eventually became upset with his American pen pal, breaking off relations after he started suspecting Loyd of stealing his ideas. Dudeney contributed to the *Strand Magazine* for over 30 years, and he wrote a number of truly challenging puzzle books that have remained staples of recreational mathematics and logic to this day. His masterpiece is considered to be his *Amusements in Mathematics* (1917).

In their 1986 book, *The Liar*, mathematician Jon Barwise and philosopher John Etchemendy approach the Liar Paradox outside of strict logic. They maintain that the paradox arises only because it is not tied to real-life contexts. So, for instance, when Epimenides says "All Cretans are liars," he may be doing so simply to confound his interlocutors. His statement may also be the result of a confusing thought he may have had. Whatever the case, the intent of Epimenides's statement can only be determined by assessing the context in which it was uttered along with Epimenides' reasons for saying it. Once such factors are determined, no paradox arises. In the 1960s, an attempt was made by logician Lofti Zadeh (1965) to incorporate the pragmatic aspect of statements directly into logic. Zadeh claimed that the logic that could handle such matters would classify Epimenides's statement as a "half truth" or a "half falsehood," depending on the context, that is, as "true under some conditions," but "false under others."

The same paradox has actually fascinated people across time and cultures, suggesting that it is an archetype of sorts (Matilal 1990). It was once even discussed by St. Jerome (c. 342–420) in a sermon related to King David (Hunter 2000):

> I said in my alarm, Every man is a liar! Is David telling the truth or is he lying? If it is true that every man is a liar, and David's statement, "Every man is a liar" is true, then David also is lying; he, too, is a man. But if he, too, is lying, his statement that "Every man is a liar," consequently

is not true. Whatever way you turn the proposition, the conclusion is a contradiction. Since David himself is a man, it follows that he also is lying; but if he is lying because every man is a liar, his lying is of a different sort.

Aware of the dangers that paradoxes posed to logical systems, British philosophers Bertrand Russell and Alfred North Whitehead collaborated to produce a system in 1913, which they believed would be impervious to paradoxes. Russell himself used a version of the Liar Paradox, called the *Barber Paradox*, to examine the issue of logical circularity more concretely:

> The village barber shaves all and only those villagers who do not shave themselves. So, shall he shave himself?

Let us assume that the barber decides to shave himself. If he does, then he would end up being shaved, of course, but the person he would have shaved is himself—a member of the village. And this contravenes the requirement that the barber should shave "all and only those villagers who do not shave themselves." The barber has, in effect, just shaved someone in the village who shaves himself. So, let us assume that the barber decides not to shave himself. But, then, he would end up being an unshaven villager. Again this goes contrary to the stipulation that he, the barber, must shave "all and only those villagers who do not shave themselves"—which would include himself. It is not possible, therefore, for the barber to decide whether or not to shave himself.

Russell argued that such undecidability arises because the barber is a member of the village. If the barber were from a different village, the paradox would not arise. Like German philosopher Gottlob Frege (1879) before him, Russell sought to find a system of logic that would exclude self-reference, circularity, and undecidability. Using a notion developed two millennia earlier by Chrysippus of Soli, Frege had claimed that circularity could be avoided by considering the *form* of a proposition separately from any real-world *content*. In this way, one could examine the consistency of propositions, without having them correspond to anything (such as barbers and Cretans). Frege's approach was developed further by Cambridge logician Ludwig Wittgenstein (1922), who ended up much later, however, believing that mathematicians expected way too much from logic.

As mentioned (Chapter 2), in 1931 Kurt Gödel showed why such issues cannot be resolved, because a statement such as the Barber's Paradox will be found in any logical system. Before Gödel, it was taken for granted that every proposition within a mathematical system could be either proved or disproved within that system. This is exactly what Euclid did in his *Elements*. But, as we saw, Gödel showed that in a mathematical system, there will emerge a proposition which is not provable within that system. The late American logician and puzzle-maker, Raymond Smullyan (1997), provides a clever puzzle version of Gödel's argument as follows:

> Let us define a logician to be accurate if everything he can prove is true; he never proves anything false. One day, an accurate logician visited the Island of Knights and Knaves, in which each inhabitant is either a knight or a knave, and knights make only true statements and knaves make only false ones. The logician met a native who made a statement from which it follows that the native must be a knight, but the logician can never prove that he is! What was the statement?

The statement was: *"You cannot prove that I am a knight."*

- If the native were a mendacious knave, then the statement would, of course, be false. Its opposite would be true—namely, *"You can prove that I am a knight."* But the puzzle asserts that an accurate logician is incapable of proving anything false. So, he cannot prove the uttered falsehood.
- If the native were a knight, then the statement *"You cannot prove that I am a knight"* would be true. But the logician cannot prove the native is a knight—the statement declares as much. So, even though the native is a knight, the logician will never be able to prove it.

An interesting demonstration of the same type of logical circularity is the one by British mathematician Philip Jourdain, which he devised in 1913:

> The following is printed on one side of a card: "The statement on the other side of this card is true." But on the card's other side the statement reads: "The statement on the other side of this card is false." What do you make of the card?

The card makes us go back and forth, from side to side, without being able to figure out what to make of it. Let "A" represent "The statement on the other side of this card is true"; and "B" represent "The statement on the other side of this card is false." It follows that:

- If A is true, then so is B, since A declares it. But if so, then A is actually false, because B, being true, states that A is false, which is a contradiction.
- If A is false, then B is false, too. But if B is false, then its statement is true, which is a contradiction.

Carroll's Take on Logic

As a teacher of mathematics and a lover of paradoxes, Lewis Carroll designed many logic puzzles to both train his students in systematic reasoning and to entertain them at the same time, in the same spirit as Alcuin. Below is one of Carroll's bewildering logic puzzles (from Carroll 1896):
 What can you conclude from the following statements?

1. Babies are illogical.
2. Nobody is despised who can manage a crocodile.
3. Illogical persons are despised.

We can examine this puzzle using sets. First, we label the relevant sets: *D* = persons who are despised; *Not-D* = persons who are not despised; *I* = illogical persons; *B* = babies; *C* = persons who can manage a crocodile. The sets *D* and *Not-D* hold no common elements, because persons belong to one or the other, but not to both. They can be shown as two squares, unconnected to each other (Figure 6.2).
 I is a subset of *D*, because statement (3) asserts that *illogical* persons are *despised*. This is shown with a circle placed inside the *D*-square (Figure 6.2). Statement (1) says that babies are illogical. This means that *B* is itself a subset of *I*. This is shown with a smaller circle inscribed within the larger *I* circle inside the *D*-square (Figure 6.2). Finally, the set *C*, which consists of persons who can manage a crocodile, is a subset of *Not-D* (= persons who are not despised), as statement (2) tells us. *C* can thus be represented by a circle inscribed within the *Not-D* square (Figure 6.2). In this way, we can show the logic involved in the statements, even though they are pure nonsense. Carroll showed, in effect, that logic does not have to make sense; it just has to be consistent.

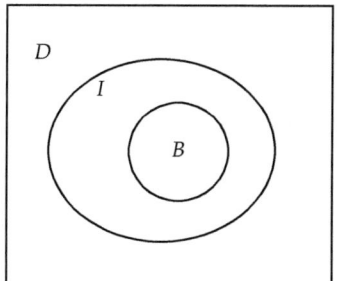 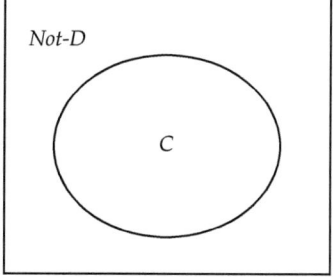

Figure 6.2 Carroll's Logic Puzzle

Recreational Logic

Puzzles that require the solver to draw conclusions from a series of statements, to figure out certain relations (such as kinship ones), to identify which element in a set does not belong, and so on are part of a genre within recreational mathematics called, specifically, *recreational logic*. The starting point for this genre is, as mentioned, Alcuin's kinship problem. Alcuin was likely aiming to get his readers to engage in drawing inferences from certain relations, and reaching the only possible conclusion from this process.

Among the myriad puzzles in this genre, the one below by Sam Loyd (1914) is one of the best-known ones:

> Uncle Reuben was in the big city to visit his sister, Mary Ann. They were walking together along a city street when they came to a small hotel. "Before we go any farther," Reuben said to his sister, "I should like to stop a moment and inquire about a sick nephew of mine who lives in this hotel." "Well," replied Mary Ann, "seeing as I don't happen to have any sick nephew to worry about, I will just trot on home. We can continue our sightseeing this afternoon." What relation was Mary Ann to the mysterious nephew?

The answer to Loyd's conundrum is that she was the sick boy's mother. This is a trick puzzle—its phraseology is such that it deflects attention away from Mary Ann as the mother (Gardner 1982). Alcuin's Problem 14 is a trick puzzle, and may also be the first of its kind. Trick puzzles have become standard in collections of recreational logic problems. Here is an illustrative one, found commonly in such anthologies:

> If you saw three shadows on three fence posts, one painted white, one painted blue and one painted red, which shadow would be the darkest?

A *shadow* does not have degrees of darkness—it casts the same shade wherever it occurs. One of the most famous of all such puzzles, which presents information in such a way that it dupes us into believing that there is a missing dollar, is the following one, which is traced to the 1930s (Abraham 1933, Read 1933), but which may go as far back as the late eighteenth century (Darling 2004, Singmaster 2004):

> Three women decide to go on a holiday to Las Vegas. They share a room at a hotel which is charging 1920s rates as a promotional gimmick. The women are charged only $10 each, or $30 in all. After going through his

guest list, the manager discovers that he has made a mistake and has actually overcharged the three vacationers. The room in which the three women are staying costs only $25. So, he gives a bellhop $5 to return to them. The sneaky bellhop knows that he cannot divide $5 into three equal amounts. Therefore, he pockets $2 for himself and returns only $1 to each woman. Now, here's the conundrum. Each woman paid $10 originally and got back $1. So, in fact, each woman paid $9 for the room. The three of them together thus paid $9 × 3, or $27 in total. If we add this amount to the $2 that the bellhop dishonestly pocketed, we get a total of $29. Yet the women paid out $30 originally! Where is the other dollar?

The trap in this puzzle is not to be found in any single word, but in the way in which the facts are laid out. Here is how these should be interpreted in order to avoid the apparent discrepancy. Originally, the women paid out $30 for the room. That is how much money was in the hands of the hotel manager when he realized that he had overcharged them. He kept $25 of the $30, and gave $5 to the bellhop to return to the women. Each woman got back $1. This means that each one paid $9 for a room. Thus, altogether the three women spent $27. Of this money, the hotel got $25 and the other $2 was pilfered by the devious bellhop. So, there is no missing dollar.

Another way to explain this puzzle is as follows. We start by noting that the women paid out $30 for the room. Of this money, the manager kept $25. The women got back $3 ($1 each). So far this adds up to $25 + $3 = $28. The remaining $2 were pocketed by the bellhop. Again, there is no missing dollar.

Epilogue

The twentieth-century philosopher Ludwig Wittgenstein (1922) once wrote that: "Logic takes care of itself; all we have to do is to look and see how it does it." These words can be used to provide a rationale behind the inclusion of two seemingly trivial problems in the *Propositiones*. Since it is impossible to define logic, because it is not a single construct, all we can do practically is to observe how it works as we use it in solving tricky puzzles—hence their use in Alcuin's text.

One of the inherent features of the two puzzles is that they demonstrate that the thinking involved is hardly haphazard, or based on trial and error. On the contrary, the given facts must be examined in terms of how they relate to each other and to reality. This shows that logical inference intersects considerably with ingenuity and common sense. Since

Alcuin's problems constitute the first examples of puzzles in recreational logic, a branch of recreational mathematics, they clearly have historical importance.

Explorations

Most of the explorations are kinship-type problems, a number of which are classic ones in recreational logic.

1. *A Take on a Conundrum by Dudeney*

Two ladies are conversing with each other, discussing a certain gentleman. The first lady asks the second one: "How is he related to you, dear?" The second lady answers: "The gentleman's mother is my mother's mother as well." How is the gentleman related to the second lady?

[This puzzle is a modified version of a problem from Dudeney's *Amusement in Mathematics* (1917).]

2. *An Enigmatic Portrait*

A painter decides to draw the portrait of a young man. As she is painting, she thinks to herself. "That boy's father is my son-in-law." Who is the painter?

3. *Another Enigmatic Portrait*

A man is looking at the portrait of a boy, uttering to himself, "Brothers and sisters have I none, but that boy's father is my father's son." Who is the boy in the portrait?

[This puzzle was invented by the late Raymond Smullyan in his 1978 book, *What Is the Name of This Book*? It is now included in virtually all anthologies of recreational logic.]

4. *A Take on Smullyan's Puzzle*

A woman is looking at a portrait of a woman, thinking to herself, "I am an only child, yet that woman's father is my father as well." Who is the woman in the portrait?

5. *Yet Another Portrait Enigma*

A young man, who is an only child, looks at the portrait of a woman, thinking to himself, "That woman's son is my mother's only child." Who is the woman?

6. *One More Portrait Enigma*

The other day, I saw a man looking at a portrait of his widow's sister. Who was the man? Is this even possible?

[This is actually a take on a well-known puzzle in trick logic.]

7. *A Family Relationship*

Two young women are discussing their families. One of the two asks the other, "How are we related?" The second woman, being a puzzle lover, answered as follows: "Your mother's son is my cousin." How are they related?

8. *Is It Possible?*

Is it possible for a woman to be the actual daughter of another woman's husband?

9. *How Many Are There?*

At a family reunion, there were three mothers, two aunts, three siblings, and three cousins. Yet the actual number of people at the reunion was 6. How so?

[This is a take on a famous puzzle by Dudeney in his *Amusements in Mathematics* (1917).]

10. *A Take on the Missing Dollar Conundrum*

Yesterday, in a suburban mall, the first customer in a computer store gave the salesclerk a $10 bill for a $3 device. The salesclerk, having no change, took the $10 bill across the corridor to the clothing store to get it broken down into ten $1 bills. The salesclerk then gave the customer the device worth $3 and seven $1 bills as change. An hour later the clothing salesclerk brought back the $10 bill demanding her money back from the bookstore salesclerk, claiming that the bill was counterfeit. To avoid quarreling, the computer salesclerk decided to give her ten $1 bills, taking back the counterfeit bill. That means that the computer salesclerk was out $3 (= cost of the book), plus the $10 bills he gave to the clothing salesclerk. Altogether he lost $13. But only $10 were used in the whole transaction. What happened?

[This is a well-known version of the missing dollar puzzle.]

Cited Works and Further Reading

Abraham, R. M. (1933). *Diversions and Pastimes*. New York: Dover.

Barwise, Jon and Etchemendy, John (1986). *The Liar*. Oxford: Oxford University Press.

Burkholder, Peter (1993). Alcuin of York's *Propositiones ad acuendos juvenes*: Introduction, Commentary & Translation. *History of Science & Technology Bulletin*, Vol. 1, number 2.

Carroll, Lewis (1896). *Symbolic Logic*. London: Macmillan.

Codd, Edgar F (1969), *Derivability, Redundancy, and Consistency of Relations Stored in Large Data Banks*, Research Report, IBM.

Darling, David J. (2004). *The Universal Book of Mathematics: From Abracadabra to Zeno's Paradoxes*. Hoboken, N.J.: Wiley.

Davis, Phillip J. and Hersh, Reuben (1986). *Descartes' Dream: The World according to Mathematics*. Boston: Houghton Mifflin.

Descartes, René (1637). *Essaies philosophiques*. Leyden: L'imprimerie de Ian Maire.

Dudeney, Henry E. (1917). *Amusements in Mathematics*. London: Thomas Nelson and Sons.

Dudeney, Henry E. (1919). *The Canterbury Puzzles*. London: Thomas Nelson and Sons.

Frege, Gottlob (1879). *Begiffsschrift eine der aritmetischen nachgebildete Formelsprache des reinen Denkens*. Halle: Nebert.

Gardner, Martin (1982). *Gotcha! Paradoxes to Puzzle and Delight*. San Francisco: Freeman.

Gödel, Kurt (1931). Über formal unentscheidbare Sätze der Principia Mathematica und verwandter Systeme, Teil I. *Monatshefte für Mathematik und Physik* 38: 173–189.

Hunter, David G. (2000). The Virgin, the Bride, and the Church: Reading Psalm 45 in Ambrose, Jerome, and Augustine. *Church History* 69: 281–303.

Jourdain, Philip (1913). The Nature and Validity of the Principle of Least Action. *The Monist* 23: 277–293.

Leibniz, Gottfried Wilhelm (1690). *De Arte Combinatoria*. Berlin: Akademie Verlag, 1923.

Loyd, Sam (1914). *Cyclopedia of Tricks and Puzzles*, New York: Dover.

Matilal, B. K. (1990). *The Word and the World: India's Contribution to the Study of Language*. Oxford: Oxford University Press.

O'Connor, J. J. and Robertson, E. F. (2012). *Propositiones ad acuendos juvenes*. https://mathshistory.st-andrews.ac.uk/HistTopics/Alcuin_book/.

Olivastro, Dominic (1993). *Ancient Puzzles: Classic Brainteasers and Other Timeless Mathematical Games of the Last 10 Centuries*. New York: Bantam.

Peirce, Charles S. (1931–1958). *Collected Papers*. Cambridge, Mass.: Harvard University Press.

Read, Cecil B. (1933). Mathematical Fallacies. *School Science and Mathematics* 33: Whole Issue.

Russell, Bertrand and Alfred N. Whitehead (1913). *Principia Mathematica*. Cambridge: Cambridge University Press.

Singmaster, David (2004). Missing Dollar and Other Erroneous Accounting. *Sources in Recreational Mathematics: An Annotated Bibliography*, https://www.puzzlemuseum.com/singma/singma6/Sources/singma-sources-edn8-2004-03-19.htm.

Smullyan, Raymond (1978). *What Is the Name of This Book?* New York: Touchstone Books.

Smullyan, Raymond (1997). *The Riddle of Scheherazade and Other Amazing Puzzles, Ancient and Modern.* New York: Knopf.

Turing, Alan (1936). On Computable Numbers with an Application to the Entscheidungs Problem. *Proceedings of the London Mathematical Society* 42: 230–265.

Wallwork, Adrian (1997). *Discussions A-Z.* Cambridge: Cambridge University Press.

Wittgenstein, Ludwig (1922). *Tractatus Logico-Philosophicus.* London: Routledge and Kegan Paul.

Zadeh, Lofti (1965). Fuzzy Sets. *Information and Control* 8: 338–353.

7
River Crossing Problems

Prologue

This book began with a brief discussion of what is Alcuin's most famous problem of all—one of his river crossing problems (Introduction). It has become a prototype for what recreational mathematics is about, differing from most of his other problems, since it deals in a playful way with combinatorics—the mathematics of combinations. There are actually four river crossing problems, which bring out different aspects of combinatorial structure as well as how to connect elements in a graph, thus constituting one of the first examples of what graph theory would come to entail centuries later.

River crossing problems constitute examples of recurring mathematical concepts that appear in other languages and at other time eras, but with the same structural blueprint (Ascher 1990). The situations that river crossing problems portray might seem obvious in a practical sense, but it is their abstract mathematical structure that has had many implications, especially in the fields of combinatorics and graph theory, as mentioned. Versions of Alcuin's river crossing problems have also been used as model problems for teaching elementary computer programming.

The Problems

The four river crossing puzzles in the *Propositiones* are numbers 17, 18, 19, and 20. Note that these have been updated linguistically here, as well as completed, since a few were missing parts.

17. *Problem of Three Men and Their Sisters*

There were three brothers, each having an unmarried sister, who wanted to cross a river. Coming to the river, they found only a small boat in which only two persons could cross at a time. How did they cross the river, so that none of the women was ever left alone with a man other than her brother either in the boat or on a bank? What is the least number of trips that are required?

18. *Problem of the Wolf, the Goat, and the Cabbage*

A traveler comes to a riverbank with a wolf, a goat, and a head of cabbage. He sees a boat that he can use for crossing over to the other bank, but to his dismay, he notices that it can carry no more than two—the traveler himself and just one other. As the traveler knows, if left alone together on either bank, the goat will eat the cabbage and the wolf will eat the goat. The wolf does not eat cabbage. How does the traveler transport his animals and his cabbage to the other side intact in a minimum number of trips?

[This is the classic one in the set, as discussed in the introduction. The question at the end is added to complete the problem.]

19. *Problem of a Man, His Wife, and Two Children*

A man and his wife, each the weight of a loaded cart, who had two children, each the weight of a small cart, needed to cross a river. However, the boat they came across could only carry the weight of one cart. Devise a way in which the crossing could be made so that the boat would not sink in the least number of trips.

[In his own solution, Alcuin refers to a cart as a weight measure. Interpreting his solution, therefore, the problem can be viewed as saying that a boat can carry at maximum only one adult or two children at once.]

20. *Problem of a River Crossing*

A man and woman and two children weighing . . . wished to cross a river.

[This problem is clearly incomplete. So, it can only be inferred as to what the complete one was supposed to be. Given that Alcuin often repeats the same kind of problem in various ways, as we have seen previously, it can be assumed that this problem is similar to the previous one, as O'Connor and Robertson (2012) also suggest.]

Solutions

17. *Answer*: nine trips

Solution

Like many medieval problems, the wording of Problem 17 provides a cultural snapshot of how family (and gender) relations were perceived in the medieval era—today, the way it portrays these relations would seem antiquated in some cultures. Here's how the crossings can be achieved under the given restrictions.

1. A brother and his sister cross over.
2. The brother stays on the other side, and the sister goes back alone.

3. Once back, she picks up one of the two other sisters and they cross over together.
4. On the other side, the first sister gets off and stays with her brother there, while the second one goes back alone.
5. Back on the original side, she picks up her own brother, and the two cross over together.
6. Once on the other side, she drops off her brother there to stay with the other brother–sister pair. She then goes back alone.
7. Back on the original side, she picks up the last of the sisters, and the two cross over together.
8. On the other side, the second sister gets off to stay with her own brother and the other brother–sister pair, while the third sister goes back alone.
9. On the original side the third sister picks up her brother, the last person there, and the two cross over together. Once they reach the other side, all three brother-sister pairs can continue with their journey.

The whole scene can be represented graphically. Let B_1, B_2, B_3 stand for each brother and S_1, S_2, S_3 for their respective sisters. Below is the graph showing how the crossings described above are made—note that "0" refers to the initial and end state (Figure 7.1).

Original Side	Boat	Other Side
0. $B_1 S_1 B_2 S_2 B_3 S_3$	$--$	$------$
1. $\_\_ B_2 S_2 B_3 S_3$	$B_1 S_1 \rightarrow$	$------$
2. $\_\_ B_2 S_2 B_3 S_3$	$\leftarrow \_ S_1$	$B_1 \_\_\_\_\_$
3. $\_\_ B_2 \_ B_3 S_3$	$S_1 S_2 \rightarrow$	$B_1 \_\_\_\_\_$
4. $\_\_ B_2 \_ B_3 S_3$	$\leftarrow \_ S_2$	$B_1 S_1 \_\_\_\_$
5. $\_\_\_\_ B_3 S_3$	$B_2 S_2 \rightarrow$	$B_1 S_1 \_\_\_\_$
6. $\_\_\_\_ B_3 S_3$	$\leftarrow \_ S_2$	$S_1 B_1 B_2 \_\_\_$
7. $\_\_\_\_ B_3 \_$	$S_2 S_3 \_ \rightarrow$	$S_1 B_1 B_2 \_\_\_$
8. $\_\_\_\_ B_3 \_$	$\leftarrow \_ S_3$	$S_1 B_1 B_2 S_2 \_\_$
9. $_____$	$B_3 S_3 \rightarrow$	$B_1 S_1 B_2 \_ B_3 \_$
0. $_____$	$--$	$B_1 S_1 B_2 S_2 B_3 S_3$

Figure 7.1 Graphic Model for Problem 17

18. *Answer*: seven trips

Solution

The solution hinges on the traveler making the first trip over successfully. Obviously, the traveler cannot start with the cabbage, since the wolf would eat the goat; nor the wolf, since the goat would then eat the cabbage. So, his only choice is to start with the goat. Once this critical decision is made, the rest of the problem is solved easily. A breakdown of the trips is as follows:

1. The traveler takes his first trip across the river to the other bank taking the goat with him, leaving the wolf and cabbage safely alone on the original side.
2. When he is on the other bank, he drops off the goat, and he goes back alone to the original side, for his second trip.
3. On that side, he can then pick up either one—the wolf or the cabbage. It does not matter, since the final result (the number of trips) will not change. So, let's go with the wolf. The traveler now goes across to the other bank with the wolf on board for his third trip.
4. On the other side, he drops off the wolf, but goes back with the goat, otherwise the wolf would eat the goat. So, on his fourth trip—back to the original side—he has the goat as a passenger.
5. Once back on the original side, he drops off the goat, picks up the cabbage and with it embarks on his fifth trip to the other bank.
6. When there, he drops off the cabbage, leaving it safely with the wolf. He then goes back to the original side alone. This is his sixth trip across the river.
7. Back on the original side, he picks up the goat and goes over to the other side with it. When he gets there, he has the wolf, goat, and cabbage, intact, and thus he can continue with his journey.

Below is the graphic model of the solution, where T = traveler, G = goat, W = wolf, and C = cabbage (Figure 7.2).

Original Side	Boat	Other Side
0. W G C (T)	_ _	_ _ _
1. W _ C	T G→	_ _ _
2. W _ C	←_ T	_ G _
3. _ _ C	T W→	_ G _
4. _ _ C	← T G	W _ _
5. _ G _	T C→	W _ _
6. _ G _	←_ T	W _ C
7. _ _ _	T G _→	W _ C
0. _ _ _	_ _	W G C (T)

Figure 7.2 Graphic Model for Problem 18

19. *Answer*: nine trips

Solution

The solution can be broken down as follows:

1. The two children get in the boat and cross over (since two children equal the weight of one adult).
2. On the other side, one of them gets off and the other one goes back alone.

3. Back on the original side, the child gets off, and the mother gets in the boat and crosses alone. It could be the father instead, but this does not change the number of trips required.
4. She gets off on the other side, and the child there brings the boat back alone.
5. Back on the original side, the child there gets in the boat with the sibling and both cross over.
6. One of them gets off on the other side and the other one brings the boat back to the waiting father.
7. The child gets off at the original side, and the father crosses over alone.
8. The father gets off at the other bank, and the child there goes back alone to get the other child.
9. Once there, the two children get on and cross over together. When they reach the other side the whole family is reunited.

Below is a graphic model of the crossings (F = father, M = mother, C_1 = one child, C_2 = the other child) (Figure 7.3).

Original Side	Boat	Other Side
0. F M C_1 C_2	$--$	$----$
1. F M $_-$ $_-$	C_1 C_2→	$----$
2. F M $_-$ $_-$	←$_-$ C_1	$---$ C_2
3. F $_-$ C_1 $_-$	M $_-$→	$---$ C_2
4. F $_-$ C_1 $_-$	←C_2 $_-$	$_-$ M $_-$ $_-$
5. F $_-$ $_-$ $_-$	C_1 C_2→	$_-$ M $_-$ $_-$
6. F $_-$ $_-$ $_-$	←C_2 $_-$	$_-$ M $_-$ C_1
7. $_-$ $_-$ $_-$ C_2	F $_-$→	$_-$ M $_-$ C_1
8. $_-$ $_-$ $_-$ C_2	←C_1 $_-$	F M $_-$ $_-$
9. $_-$ $_-$ $_-$ $_-$	C_1 C_2→	F M $_-$ $_-$
0. $_-$ $_-$ $_-$ $_-$	$--$	F M C_1 C_2

Figure 7.3 Graphic Model for Problem 19

20. *Answer*: nine trips

Solution

Since it can be assumed that this problem is similar to the previous one, at least in theory, only the graphic solution is given here for convenience: W = woman, M = man, C_1 = one child, C_2 = the other child (Figure 7.4). Note that there are slightly different possibilities, such as M and W reversing each other's trips, but the end result will be the same (Figure 7.4).

Original Side	Boat	Other Side
0. W M C_1 C_2	– –	– – – –
1. W M _ _	C_1C_2→	– – – –
2. W M _ _	←_ C_1	_ _ _ C_2
3. W _ C_1_	M _ →	_ _ _ C_2
4. W _ C_1_	← C_2-	_ M _ _
5. W _ _ _	C_1 C_2→	_ M _ _
6. W _ _ _	← C_2_	_ M _ C_1
7. _ _ _ C_2	W _ →	_ M _ C_1
8. _ _ _ C_2	← C_1_	W M _ _
9. _ _ _ _	C_1 C_2→	W M _ _
0. _ _ _ _	– –	W M C_1 C_2

Figure 7.4 Graphic Model for (Hypothetical) Problem 20

Annotations

The river crossing problems have had a number of important implications for mathematics subsequent to Alcuin; only a few can be mentioned here. They are among the earliest treatments of combinatorics—a branch of mathematics that studies the optimal ways to make permutations and combinations of elements under given constraints. Although the study of combinatorics is implicit in various ancient mathematical texts, it was never really formulated as a distinct mode of inquiry before Alcuin's river crossing problems. Also, since the number of trips across in the problems is required to be minimal, the problems also embed the notion of an "optimal path," or the analysis of the optimal or possible sequence of moves needed for an operation, which is central to graph theory.

Versions

Alcuin's river crossing problems piqued the interest of many subsequent mathematicians, not to mention puzzle-makers. In his remarkable work of theoretical and recreational mathematics, *De viribus quantitates* (1500), Italian mathematician Luca Pacioli discussed a version of Alcuin's Problem 17, in which a boat is allowed to hold more than two people; and in his *Trattato* of 1556, Niccolò Tartaglia discussed a three-couple version, in which three brides and their jealous husbands had to go across the river— yet another culturally revealing wording of a math problem. In this case, the stipulation is that no wife can be left on either side or on the boat without the presence of her husband. Wives can be alone together on the boat (without any husband).

One possible solution is given below. Note that H_1, H_2, H_3, W_1, W_2, and W_3 stand for the three husbands and their three wives, respectively; that is, H_1 is the husband of W_1, H_2 is the husband of W_2, and H_3 is the husband of W_3 (Figure 7.5).

Original Side	Boat	Other Side
0. $H_1 W_1 H_2 W_2 H_3 W_3$	$\_\ \_$	$\_\ \_\ \_\ \_\ \_\ \_$
1. $\_\ \_ H_2 W_2 H_3 W_3$	$H_1 W_1 \rightarrow$	$\_\ \_\ \_\ \_\ \_\ \_$
2. $\_\ \_ H_2 W_2 H_3 W_3$	$\leftarrow \_ W_1$	$H_1 \_\ \_\ \_\ \_\ \_$
3. $\_\ \_ H_2 \_ H_3 W_3$	$W_1 W_2 \rightarrow$	$H_1 \_\ \_\ \_\ \_\ \_$
4. $\_\ \_ H_2 \_ H_3 W_3$	$\leftarrow \_ W_2$	$H_1 W_1 \_\ \_\ \_\ \_$
5. $\_\ \_\ \_\ \_ H_3 W_3$	$H_2 W_2 \rightarrow$	$H_1 W_1 \_\ \_\ \_\ \_$
6. $\_\ \_\ \_\ \_ H_3 W_3$	$\leftarrow \_ W_2$	$H_1 W_1 H_2 \_\ \_\ \_$
7. $\_\ \_\ \_\ \_ H_3 \_$	$W_2 W_3 \rightarrow$	$H_1 W_1 H_2 \_\ \_\ \_$
8. $\_\ \_\ \_\ \_ H_3 \_$	$\leftarrow \_ W_3$	$H_1 W_1 H_2 W_2 \_\ \_$
9. $\_\ \_\ \_\ \_\ \_\ \_$	$H_3 W_3 \rightarrow$	$H_1 W_1 H_2 W_2 \_\ \_$
0. $\_\ \_\ \_\ \_\ \_\ \_$	$\_\ \_$	$H_1 W_1 H_2 W_2 H_3 W_3$

Figure 7.5 Graphic Model for Tartaglia's Puzzle

A nineteenth-century version involves three missionaries and three cannibals—constituting yet another cultural version that would be considered odd today, not to mention illegal. The missionaries and cannibals are together at the start on the original side. Cannibals must never be allowed to outnumber missionaries on either side. A missionary and a cannibal can be on the boat together, which can only hold two. How can they all get across safely (Pressman and Singmaster 1989)? Here is one possible solution: M_1, M_2, and M_3 stand for the missionaries and C_1, C_2, and C_3 stand for the cannibals (Figure 7.6).

Original Side	Boat	Other Side
0. $M_1 M_2 M_3 C_1 C_2 C_3$	$\_\ \_$	$\_\ \_\ \_\ \_\ \_\ \_$
1. $\_ M_2 M_3 \_ C_2 C_3$	$M_1 C_1 \rightarrow$	$\_\ \_\ \_\ \_\ \_\ \_$
2. $\_ M_2 M_3 \_ C_2 C_3$	$\leftarrow M_1 \_$	$\_\ \_\ \_ C_1 \_\ \_$
3. $\_ M_2 M_3 \_\ \_ C_3$	$M_1 C_2 \rightarrow$	$\_\ \_\ \_ C_1 \_\ \_$
4. $\_ M_2 M_3 \_\ \_ C_3$	$\leftarrow C_2 \_$	$M_1 \_\ \_ C_1 \_\ \_$
5. $\_\ \_ M_3 \_\ \_ C_3$	$M_2 C_2 \rightarrow$	$M_1 \_\ \_ C_1 \_\ \_$
6. $\_\ \_ M_3 \_\ \_ C_3$	$\leftarrow C_2$	$M_1 M_2 \_\ C_1 \_\ \_$
7. $\_\ \_\ \_\ \_\ \_ C_3$	$M_3 C_2 \rightarrow$	$M_1 M_2 \_\ C_1 \_\ \_$
8. $\_\ \_\ \_\ \_\ \_ C_3$	$\leftarrow C_2$	$M_1 M_2 M_3 C_1 \_\ \_$
9. $\_\ \_\ \_\ \_\ \_\ \_$	$C_2 C_3 \rightarrow$	$M_1 M_2 M_3 C_1 \_\ \_$
0. $\_\ \_\ \_\ \_\ \_\ \_$	$\_\ \_$	$M_1 M_2 M_3 C_1 C_2 C_3$

Figure 7.6 Graphic Model for the Missionaries and Cannibals Puzzle

As examples such as these show the details might vary, but the underlying mathematical blueprint is the same. As Martha Ascher (1990: 26) has aptly observed, the different cultural versions of the puzzle "are expressions of their cultures and so variations will be seen in the characters, the settings, and the way in which the logical problem is framed." An interesting version, also traced to the nineteenth century, is paraphrased below:

> Three soldiers have to cross a river without a bridge. Two boys with a boat agree to help the soldiers, but the boat is so small it can support only one soldier or two boys at one time. So, a soldier and a boy cannot be in the boat at the same time for fear of sinking it. Given that none of the soldiers can swim, it would seem that under these circumstances just one soldier could cross the river. Yet all three soldiers eventually end up on the other bank and return the boat to the boys. How do they do it?

It takes the following crossings: S_1, S_2, and S_3 stand for the soldiers and B_1 and B_2 stand for the boys. Note that there may be other ways to solve the problem (Figure 7.7).

Original Side	Boat	Other Side
0. $S_1\,S_2\,S_3\,B_1\,B_2$	$-\ -$	$-\,-\,-\,-\,-$
1. $S_1\,S_2\,S_3\,\_\,\_$	$B_1\,B_2\,\_ \rightarrow$	$-\,-\,-\,-\,-$
2. $S_1\,S_2\,S_3\,\_\,\_$	$\leftarrow B_1\,\_$	$\_\,\_\,\_\,\_\,B_2$
3. $\_\,S_2\,S_3\,B_1\,\_$	$S_1\,\_ \rightarrow$	$\_\,\_\,\_\,\_\,B_2$
4. $\_\,S_2\,S_3\,B_1\,\_$	$\leftarrow B_2\,\_$	$S_1\,\_\,\_\,\_\,\_$
5. $\_\,S_2\,S_3\,\_\,\_$	$B_1\,B_2 \rightarrow$	$S_1\,\_\,\_\,\_\,\_$
6. $\_\,S_2\,S_3\,\_\,\_$	$\leftarrow B_2\,\_$	$S_1\,\_\,\_\,B_1\,\_$
7. $\_\,\_\,S_3\,\_\,B_2$	$S_2\,\_ \rightarrow$	$S_1\,\_\,\_\,B_1\,\_$
8. $\_\,\_\,S_3\,\_\,B_2$	$\leftarrow B_1\,\_$	$S_1\,S_2\,\_\,\_\,\_$
9. $\_\,\_\,S_3\,\_\,\_$	$B_1\,B_2 \rightarrow$	$S_1\,S_2\,\_\,\_\,\_$
10. $\_\,\_\,S_3\,\_\,\_$	$\leftarrow B_2\,\_$	$S_1\,S_2\,\_\,B_1\,\_$
11. $\_\,\_\,\_\,\_\,B_2$	$S_3\,\_ \rightarrow$	$S_1\,S_2\,\_\,B_1\,\_$
12. $\_\,\_\,\_\,\_\,B_2$	$\leftarrow B_1\,\_$	$S_1\,S_2\,S_3\,\_\,\_$
13. $\_\,\_\,\_\,\_\,\_$	$B_1\,B_2 \rightarrow$	$S_1\,S_2\,S_3\,\_\,\_$
0. $\_\,\_\,\_\,\_\,\_$	$-\ -$	$S_1\,S_2\,S_3\,B_1\,B_2$

Figure 7.7 Graphic Model for the Soldiers and Boys Puzzle

In his *Récréations mathématiques* of 1892, Édouard Lucas examined the situation of four couples crossing a river, which cannot be solved without the presence of an island as a temporary landing place in between the two banks of the river, elaborating upon a problem suggested by Cadet de Fontenay in 1879. Various four couple versions have been since devised (Dudeney 1917, Pressman and Singmaster 1989). Below is one formulated by American puzzle-maker Sam Loyd (1914), known as the "Four Elopements Puzzle."

The story of four elopements says that four men eloped with their sweethearts, but in carrying out their plan were compelled to cross a stream in a boat which would hold but two persons at a time. It appears that the young men were so extremely jealous that not one of them would permit his prospective bride to remain at any time in the company of any other man or men unless he was also present. Nor was any man to get into a boat alone, when there happened to be a girl alone on the island or shore, other than the one to whom he was engaged. This feature of the condition looks as if the girls were also jealous and feared that their fellows would run off with the wrong girl if they got a chance. Well, be that as it may, the problem is to guess the quickest way to get the whole party across the river according to the conditions imposed. Let us suppose the island to be in the middle of the stream. Now, tell how many minimum number of trips would the boat make to get the four couples safely across in accordance with the stipulations?

Below is a solution (Figure 7.8): M_1, M_2, M_3, M_4 and W_1, W_2, W_3, W_4 stand for the men and their brides, respectively; that is, W_1 is the prospective bride of M_1, W_2 is the prospective bride of M_2, W_3 is the prospective bride of M_3, and W_4 is the prospective bride of M_4. Note that "Boat-1" represents the boat trip before the island and "Boat-2" after the island.

Original Side	Boat-1	Island	Boat-2	Other Side
0. $M_1 M_2 M_3 M_4 W_1 W_2 W_3 W_4$	_ _	_ _ _	_ _	_ _ _ _ _ _ _ _
1. $M_1 M_2 M_3 M_4$ _ _ $W_3 W_4$	$W_1 W_2 \rightarrow$	_ _ _	$W_1 W_2 \rightarrow$	_ _ _ _ W_1 _ _ _
2. $M_1 M_2 M_3 M_4$ _ _ $W_3 W_4$	$\leftarrow W_2$ _	_ _ _	$\leftarrow W_2$ _	_ _ _ _ W_1 _ _ _
3. $M_1 M_2 M_3 M_4$ _ _ _ W_4	$W_2 W_3 \rightarrow$	_ _ _	_ _	_ _ _ _ W_1 _ _ _
4. $M_1 M_2 M_3 M_4$ _ _ _ W_4	$\leftarrow W_3$ _	W_2 _ _	_ _	_ _ _ _ W_1 _ _ _
5. _ _ $M_3 M_4$ _ _ $W_3 W_4$	$M_1 M_2 \rightarrow$	W_2 _ _	$M_1 M_2 \rightarrow$	_ _ _ _ W_1 _ _ _
6. _ _ $M_3 M_4$ _ _ $W_3 W_4$	$\leftarrow M_2$ _	W_2 _ _	$\leftarrow M_2$ _	M_1 _ _ _ W_1 _ _ _
7. _ $M_2 M_3 M_4$ _ _ _ _	$W_3 W_4 \rightarrow$	$W_2 W_3 W_4$	_ _	M_1 _ _ _ W_1 _ _ _
8. _ $M_2 M_3 M_4$ _ _ _ _	$\leftarrow W_4$ _	$W_2 W_3$ _	_ _	M_1 _ _ _ W_1 _ _ _
9. _ _ _ M_4 _ _ _ W_4	$M_2 M_3 \rightarrow$	$W_2 W_3$ _	$M_2 M_3 \rightarrow$	M_1 _ _ _ W_1 _ _ _
10. _ _ _ M_4 _ _ _ W_4	_ _	$W_2 W_3$ _	$\leftarrow W_1$	$M_1 M_2 M_3$ _ _ _ _ _
11. _ _ _ M_4 _ _ _ W_4	_ _	W_2 _ _	$W_1 W_3 \rightarrow$	$M_1 M_2 M_3$ _ _ _ _ _
12. _ _ _ M_4 _ _ _ W_4	$\leftarrow M_2$ _	W_2 _ _	$\leftarrow M_2$ _	M_1 _ M_3 _ W_1 _ W_3 _
13. _ _ _ _ _ _ _ W_4	$M_2 M_4 \rightarrow$	W_2 _ _	$M_2 M_4 \rightarrow$	M_1 _ M_3 _ W_1 _ W_3 _
14. _ _ _ _ _ _ _ W_4	_ _	W_2 _ _	$\leftarrow W_3$	$M_1 M_2 M_3 M_4 W_1$ _ _ _
15. _ _ _ _ _ _ _ W_4	_ _	_ _ _	$W_2 W_3 \rightarrow$	$M_1 M_2 M_3 M_4 W_1 W_2$ _ _
16. _ _ _ _ _ _ _ W_4	$\leftarrow W_3$	_ _ _	$\leftarrow W_3$	$M_1 M_2 M_3 M_4 W_1 W_2$ _ _
17. _ _ _ _ _ _ _ _	$W_3 W_4 \rightarrow$	_ _ _	$W_3 W_4 \rightarrow$	$M_1 M_2 M_3 M_4 W_1 W_2$ _ _
0. _ _ _ _ _ _ _ _	_ _	_ _ _	_ _	$M_1 M_2 M_3 M_4 W_1 W_2 W_3 W_4$

Figure 7.8 Graphic Model for the Four Elopements Puzzle

Combinatorics

Alcuin's river crossing problems deal with making combinations of objects according to specific constraints. The idea actually goes back to antiquity. For example, the ancient Indian medical doctor Sushruta asserted in the *Sushruta Samhita* (first millennium BCE) that 63 combinations can be made out of six different gustatory tastes, taken one at a time, two at a time, and so on. In the Middle Ages, another Indian mathematician, Mahāvīra (c. 850), provided formulas for computing permutations and combinations (Puttaswamy 2000). It was around this time that Alcuin devised his ingenious problems.

Consider the following problem as a case-in-point:

> In how many possible ways can three people named Wes, Gina, and Caroline sit in three seats, with no restrictions?

Let us label the three seats (1), (2), and (3) and the individuals W = Wes, G = Gina, and C = Caroline. For each person who occupies seat (1), either one of the remaining two can occupy seat (2). This leaves one person for seat (3). Below is a summary of all the possible arrangements of the W, G, and C in the three seats (Figure 7.9).

Seat (1)	Seat (2)	Seat (3)	Arrangement
W	G	C	WGC
W	C	G	WCG
G	W	C	GWC
G	C	W	GCW
C	W	G	CWG
C	G	W	CGW

Figure 7.9 Seating Arrangements

The number of potential arrangements—WGC, WCG, GWC, GCW, CWG, CGW—is $3 \times 2 \times 1 = 6$. This expresses in arithmetical form the fact that for each of the three individuals who occupies seat (1), two others will occupy seat (2), and the remaining one will occupy seat (3). The arrangements are called permutations. If there were four seats and four people, then there would be $4 \times 3 \times 2 \times 1 = 24$ possible arrangements. To reiterate, the first digit "4" refers to the fact that any one of the four can occupy seat (1). The second digit "3" tells us that for each of the four possibilities for seat (1) there are three ways in which seat (2) can be occupied. This produces "$4 \times 3 = 12$" permutations for the first two seats. The third digit tells us that for each of the "12" permutations, there are two ways in which seat (3) can be occupied, for a total of "4 × 3 × 2 = 24" permutations. Finally, the fourth digit indicates that for each of the "24" permutations so far there is only one way in which seat (4) can be occupied, for a total of "4 × 3 × 2 × 1 = 24" permutations.

If there were five seats to be occupied on the boat and five people to fill them, there would be "5 × 4 × 3 × 2 × 1" possible permutations; if there were six seats and six people, there would be "6 × 5 × 4 × 3 × 2 × 1" permutations; and so on. In general, for "n" places to be occupied, and "n" items to fill them, there would be "$n × (n − 1) × (n − 2) × \ldots 1$" permutations. This is known as a *factorial*, and is represented with the symbol "$n!$":

$$n! = n × (n − 1) × (n − 2) × \ldots × 1$$

Here are some examples of factorials:

$$4! = 4 × 3 × 2 × 1$$
$$5! = 5 × 4 × 3 × 2 × 1$$
$$9! = 9 × 8 × 7 × 6 × 5 × 4 × 3 × 2 × 1$$
$$11! = 11 × 10 × 9 × 8 × 7 × 6 × 5 × 4 × 3 × 2 × 1$$

Now, consider the following problem in which there is a repetition of the items in the arrangement.

How many five-digit numerals can be constructed using the digits 1, 1, 2, 3, and 4?

Here we have five digits, two of which are indistinguishable—the two "1's." This means that some arrangements will turn out to be exactly the same. These must therefore be "filtered out." To do this, let us assign a subscript to the two "1's," so as to keep track of them. The five digits, therefore, can be written as follows:

$$1_1, 1_2, 2, 3, 4$$

A total number of "120" five-digit numerals can be made, using the same reasoning above for determining the number of seating arrangements. In this case there are five items to be permuted. Therefore, using factorial notation:

$$5! = 5 × 4 × 3 × 2 × 1 = 120$$

However, some of these will be indistinguishable when the subscripts are removed, as the two examples below show:

$$1_121_234 = 12134$$
$$1_221_134 = 12134$$

Now, how many of these are there among the "120" numerals? There will be "2!" of them, because that is how many permutations there are of the two digits "1_1" and "1_2" when considered in isolation from their occurrence with the other digits in the numerals. Canceling out the indistinguishable cases involves division, in which "2!" is divided into "5!":

$$\frac{5!}{2!} = \frac{5 \times 4 \times 3 \times \cancel{(2 \times 1)}}{\cancel{(2 \times 1)}}$$

And $5 \times 4 \times 3 = 60$

We can thus make "60" distinct five-digit numerals. In general, the number of distinct permutations of "n" objects among which there are "r" indistinguishable cases is:

$$\frac{n!}{r!}$$

Now, let us consider a problem in which we have more people than spots in which to put them:

Suppose you had to choose one person to be president, one vice-president, and one secretary from a committee of 10 people. In how many ways could you fill those three positions?

In this case, "10" people can be selected for the position of president; for each of these any one of the remaining "9" can be selected for the vice-presidency; and for each of the previous pairs, we can choose from among "8" people to fill the remaining position of secretary. So, the number of total permutations in this case is "$10 \times 9 \times 8 = 720$." Notice that the product "$10 \times 9 \times 8$" is, in effect, "10!" with the last seven factors canceled from it:

$$10 \times 9 \times 8 \cancel{(\times 7 \times 6 \times 5 \times 4 \times 3 \times 2 \times 1)}$$

Therefore, to determine the number of positions to be filled, we are in effect dividing "10!" by "7!" ($= 7 \times 6 \times 5 \times 4 \times 3 \times 2 \times 1$):

$$\frac{10!}{7!} = \frac{10 \times 9 \times 8 \times \cancel{(7 \times 6 \times 5 \times 4 \times 3 \times 2 \times 1)}}{\cancel{(7 \times 6 \times 5 \times 4 \times 3 \times 2 \times 1)}}$$

And $10 \times 9 \times 8 = 720$

The denominator "7!" above can be rewritten as "(10–3)!" with the "3" standing for the number of positions to be filled. In general, if the numerator is "n!" then the denominator will be "(n–r)!" where "r" stands for the number of positions to be filled:

$$\frac{n!}{(n-r)!}$$

This formula now allows us to solve any problem that requires us to put "n" objects in "r" positions.

Let us now look at one other type of arrangement pattern. Suppose we wanted to select a three-member subcommittee from the 10 candidates, with no regard as to who fills what position. Are there also "720" ways to do this? The answer is no. For example, let us say that three of the people to be chosen for the subcommittee are named Camille, Lucia, and Raul. There are "3!" ways (3! = 3 × 2 × 1 = 6) to fill the three positions on the subcommittee (Figure 7.10).

President	VP	Secretary
↓	↓	↓
Camille	Lucia	Raul
Camille	Raul	Lucia
Lucia	Camille	Raul
Lucia	Raul	Camille
Raul	Camille	Lucia
Raul	Lucia	Camille

Figure 7.10 Three Positions Filled by Three People

When making a subcommittee, however, the order of the selections is irrelevant. It only matters that three be chosen. It is irrelevant, for instance, whether the order is "Camille, Lucia, Raul" or "Raul, Lucia, Camille." This type of arrangement is called a *combination*, rather than a *permutation*. It is defined as an arrangement that is put together with no regard to order. In the case of a three-member subcommittee, how many "3!" arrangements (in no particular order) can be made from the "720" possible selections? To determine this, we simply divide "3!" into "720":

$$\frac{720}{3!}$$

↓

$$\frac{720}{6}$$

↓

$$120$$

So, there are 120 ways to make up the subcommittee. We can now generalize this. Note that the denominator in the first number above is "$r!$"—the number of positions to be filled. Note as well that the "720" in the numerator was produced by our previous formula—namely, by "$n!/(n - r)!$." So, the general formula is the previous formula divided by "$r!$":

$$\frac{n!}{(n-r)!r!}$$

Let us now apply the formula to our example:

$$\frac{10!}{(10 - 3)! \, 3!}$$
$$\downarrow$$
$$\frac{10 \times 9 \times 8 \times (7 \times 6 \times 5 \times 4 \times 3 \times 2 \times 1)}{(7 \times 6 \times 5 \times 4 \times 3 \times 2 \times 1) \times (3 \times 2 \times 1)}$$
$$\downarrow$$
$$\frac{10 \times 9 \times 8 \times \cancel{(7 \times 6 \times 5 \times 4 \times 3 \times 2 \times 1)}}{\cancel{(7 \times 6 \times 5 \times 4 \times 3 \times 2 \times 1)} \times (3 \times 2 \times 1)}$$
$$\downarrow$$
$$\frac{10 \times 9 \times 8}{(3 \times 2 \times 1)}$$
$$\downarrow$$
$$\frac{720}{6}$$
$$\downarrow$$
$$120$$

Box 7.1 Summary

The permutation of "n" objects: $n! = n \times (n - 1) \times (n - 2) \times \ldots \times 1$
The permutation of "n" objects in which "r" cases are indistinguishable: $n!/r!$
The permutation of "n" objects taken "r" at a time: $n!/(n - r)!$
The combination of "n" objects in "r" positions: $n!/(n - r)! \, r!$

The Kirkman Schoolgirls Problem

The *Kirkman Schoolgirls Problem* is a famous problem in combinatorics, named after the notable mathematician Thomas Penyngton Kirkman (1806–1895), who posed it in 1847. It is paraphrased below:

> How can 15 girls walk in 5 rows of 3 each for 7 days so that no girl walks with any other girl in the same triplet more than once?

Let us replace the "girls" with "15" integers in order starting with "0":

$$\textit{Fifteen Girls } = \{0, 1, 2, 3, 4, 5, 6, 7, 8, 9, 10, 11, 12, 13, 14\}$$

Solutions to this problem involve placing the 15 numerals into seven sets, each one representing a day of the week. In each set, we must further distribute the numerals into five rows, of three numerals each, so that no numeral appears with two other numerals in a triplet more than once. For example, the numeral "5" can be put into seven triplets, with no repeats, as follows:

$$\{0, \underline{5}, 10\}, \{12, 14, \underline{5}\}, \{1, 2, \underline{5}\}, \{4, \underline{5}, 8\}, \{\underline{5}, 7, 13\}, \{\underline{5}, 6, 9\}, \{3, \underline{5}, 11\}$$

Below is one possible solution (Figure 7.11).

Monday			Tuesday			Wednesday			Thursday		
0	5	10	0	1	4	1	2	5	4	5	8
1	6	11	2	3	6	3	4	7	6	7	10
2	7	12	7	8	11	8	9	12	11	12	0
3	8	13	9	10	13	10	11	14	13	14	2
4	9	14	12	14	5	13	0	6	1	3	9

Friday			Saturday			Sunday		
4	6	12	10	12	3	2	4	10
5	7	13	11	13	4	3	5	11
8	10	1	14	1	7	6	8	14
9	11	2	0	2	8	7	9	0
14	0	3	5	6	9	12	13	1

Figure 7.11 A Solution to Kirkman's Schoolgirls Problem

There are seven possible solutions to the problem (Biggs 1981, Tahta 2006). Combinatorics is a fascinating branch of mathematics. One must not forget that it was Alcuin's river crossing problems that planted the intellectual seeds for it to evolve in to a branch of mathematics and computer science (Csorba, Hurkens, and Woeginger 2008).

Epilogue

At first consideration, Alcuin's river crossing problems seem to just require a simple form of logic to solve, but as we have seen in this chapter, there is much more to them than meets the eye, reflecting what may well be a conceptual mathematical archetype hidden within them. In her fascinating book, *Africa Counts: Number and Pattern in African Cultures* (1973), Claudia Zaslavsky, even mentions an age-old version told by the Kpelle children in Liberia in which a man must ferry a leopard, a goat, and a bunch of Cassava leaves across the river. It is unlikely that the children would have known of Alcuin's text, or vice versa, given the geographical and linguistic distances between the two. Nevertheless, it is astounding to find the same conceptualization of a river crossing situation.

   In effect, river crossing problems appear in various ways cross-culturally and across time (Ascher 1990). They are examples of how practical reasoning can be transformed into theoretical knowledge, which is at the core of how mathematicians think, and this may well have been Alcuin's pedagogical objective with his ingenious problems.

Explorations

The explorations are either versions of Alcuin's river crossing problems or else based on simple combinatorics.

1. *Cat Crossing*

One man, one woman, and a domestic cat come to a riverbank, wanting to cross over to the other side. As in Alcuin's problems, there is a boat that can only carry two at one time. The cat cannot be with the man on either bank or on the boat, for some strange reason, without the presence of the woman. The cat can be left alone on either bank, or can be with the woman anywhere without the presence of the man. Obviously, the cat cannot steer the boat. How do the three get across under these conditions given that either the man or the woman can steer the boat?

2. *Another Cat Crossing*

This time, two men, two women, and a domestic cat come to a riverbank aiming to cross over to the other side. Again, there is a boat that can only carry two at one time. The cat cannot be with either man on either bank or on the boat, again for some strange reason, without the presence of at least one of the women. The cat can be left alone on either bank and can be with any of the women anywhere without the presence of any man. The men and women can be alone or together in any combination on any bank or on the boat. Of course, the cat cannot steer the boat. How do the five get across under these conditions, given that any one of the men or women can steer the boat?

3. *A Cat and Dog Crossing*

This time, two children, one woman, a domestic cat, and a domestic dog come to a riverbank to cross over to the other side. Again, there is a boat that can only carry two at one time. The two animals cannot be left alone on either side, without the presence of a child or the woman, but can be with any human on the boat. Either child or the woman can be left alone on either side, and they can be together on the boat in any combination. How do the five get across under these conditions, given that either child or the woman can steer the boat?

4. *A Take on Alcuin's Problem 18*

Two travelers come to a riverbank with a wolf, a goat, and a head of cabbage. The boat can carry no more than two. As we know, if left alone together on either bank, the goat will eat the cabbage and the wolf will eat the goat. The wolf does not eat cabbage. How can the travelers transport themselves and the others to the other side intact in a minimum number of trips?

5. *A Family Crossing*

A brother, a sister, a male cousin, and a female cousin come to a riverbank wanting to cross over to the other side. Again, there is a boat that can carry only two at one time. The two siblings cannot be left alone on either side without a cousin being present, otherwise they would pick on each other. Nor can the two cousins be left alone on either side, without a sibling being present, for then they too would pick on each other. The siblings and the cousins can, however, be together in any pairing on the boat or on either side. How do the four get across under these conditions, given that all four can steer the boat?

6. *Revisiting Alcuin's Problem 18*

For Alcuin's Problem 18—the traveler with a wolf, a goat, and a head of cabbage, and a boat with only two seats—could the trips be made safely if the traveler must go back to the original side always alone after dropping off any of the three on the other side?

7. *A Take on Loyd's Version*

Determine the number of complete back-and-forth crossings needed for four husband and wife pairs if the small boat that is to take them across holds only two people, and the crossings are to be so organized that no woman shall be left with a man unless her husband is present. Any pairing of women is allowed anywhere.

[Note that this version, which is imitative of the ones discussed above, reflects a situation that would be culturally anomalous today. Clearly, even mathematical puzzles cannot be free of cultural predispositions.]

8. *House Routes*

There are three houses close to each other, A, B, and C. There are three different routes from A to B and four different routes from B to C. How many routes are there from A to C that go through B?

9. *Club Membership*

A club has 20 members. They wish to elect a president and vice-president from among the membership. How many different outcomes of the election are possible? What if only two candidates, call them Brenda and Heather, are allowed to be elected president?

10. *Vegetable Soup*

A chef wants to make soup with exactly five different vegetables. If she has 12 vegetables from which to choose, how many different soups can she make?

Cited Works and Further Reading

Ascher, Marcia (1990). A River-Crossing Problem in Cross-Cultural Perspective. *Mathematics Magazine* 63: 26–29.

Bellman, Richard (1962). Dynamic Programming and Difficult Crossing Puzzles. *Mathematics Magazine* 35: 27–29.

Biggs, N. L. (1979). The Roots of Combinatorics. *Historia Mathematica* 6: 109–136.

Biggs, N. L. (1981). T. P. Kirkman, Mathematician. *The Bulletin of the London Mathematical Society* 13: 97–120.

Bollobás, Béla (2022). *The Art of Mathematics*. Cambridge: Cambridge University Press.

Burkholder, Peter (1993). Alcuin of York's *Propositiones ad acuendos juvenes*: Introduction, Commentary & Translation. *History of Science & Technology Bulletin*, Vol. 1, number 2.

Csorba, Péter, Hurkens, Cor A. J., and Woeginger, Gerhard J. (2008). The Alcuin Number of a Graph. *Algorithms: ESA 2008, Lecture Notes in Computer Science*, vol. 5193, Springer-Verlag, pp. 320–331.

Dudeney, Henry E. (1917). *Amusements in Mathematics*. London: Thomas Nelson and Sons.

Fraley, Robert, Cooke, Kenneth L., and Detrick, Peter (1966). Graphical Solution of Difficult Crossing Puzzles. *Mathematics Magazine* 39: 151–157.

Franci, Raffaella (2002). *Jealous Husbands Crossing the River: A Problem from Alcuin to Tartaglia*. Stuttgart: Franz Steiner Verlag.

Gerdes, Paulus (1994). On Mathematics in the History of Sub-Saharan Africa. *Historia Mathematica* 21: 23–45.

Kirkman, Thomas P. (1847). On a Problem in Combinations. *The Cambridge and Dublin Mathematical Journal* 2: 191–204.

Loyd, Sam (1914). *Cyclopedia of Tricks and Puzzles*. New York: Dover.

Lucas, Édouard (1892). *Récréations mathématiques*. Paris: Albert Blanchard.

O'Connor, J. J. and Robertson, E. F. (2012). *Propositiones ad acuendos juvenes*. https://mathshistory.st-andrews.ac.uk/HistTopics/Alcuin_book/.

Pacioli, Luca (1500). *De viribus quantitatis*. Milano: Ente Raccolta Vinciana, 1997.

Peterson, Ivars (2003). Tricky Crossings. *Science News* 164: 325.

Pressman, Ian and Singmaster, David (1989). The Jealous Husbands and The Missionaries and Cannibals. *The Mathematical Gazette* 73: 73–81.

Primrose, E. J. F. (1976). Kirkman's Schoolgirls in Modern Dress. *The Mathematical Gazette* 60: 292–293.

Puttaswamy, Tumkur K. (2000). The Mathematical Accomplishments of Ancient Indian Mathematicians. In: Helaine Selin (ed.). *Mathematics Across Cultures: The History of Non-Western Mathematics*. *Netherlands*: Kluwer Academic Publishers.

Schwartz, Benjamin L. (1961). An Analytic Method for the Difficult Crossing Puzzles. *Mathematics Magazine* 34: 187–193.

Singmaster, David (2015). *Problems for Metagrobologists*. Singapore: World Scientific.

Tahta, Dick (2006). *The Fifteen Schoolgirls*. Cambridge: Black Apollo Press.

Zaslavsky, Claudia (1973). *Africa Counts: Number and Pattern in African Cultures*. Boston: Prindle, Weber, and Schmidt.

8
Sequences

Prologue

At York, Alcuin established a school and a library, developing a broad reputation as a skillful teacher and scholar, attracting the attention of Charlemagne, who invited him to his court (Introduction). It was at the court that Alcuin developed his recreational approach to teaching mathematics. He was seemingly determined to get people to reap enjoyment as they worked through his *Propositiones*, making mathematics palatable to anyone, from arithmetic and geometry to Diophantine equations. The two problems discussed in this chapter deal with yet one other area that Alcuin covered in his text—sequences. A *sequence* is as an arrangement of objects in a particular order; a *series* is a sequence in which the elements are related to each other by some mathematical operation. The importance of sequences to mathematics cannot be overestimated. From ancient paradoxes dealing with infinity to the modern theory of sets, this notion is a central one in mathematics.

Interestingly, Problem 42 may be the inspiration for the development of the formula for the sum of the numbers in a series, prefiguring Carl Friedrich Gauss (discussed below). These two problems add to the significance of the *Propositiones* to mathematics and how much its evolution may depend on simple problems that bring forth some new idea within them.

The Problems

The two relevant problems in the *Propositiones* are numbers 13 and 42. A third one—number 12—is sometimes included under the same rubric (Bindner and Erickson 2012). But since it deals primarily with measurement, it will be discussed in the next chapter.

13. *Problem of the King's Army*

A king ordered his servant to assemble an army from 30 villages as follows. He should bring back from each successive village as many men as he had taken there. So, the servant went to the first village alone; he went with one other man to the second village; he went with the men he had so far conscripted to the third village; and so on. How many men were conscripted from the 30 villages?

42. *Problem of the Pigeons and the Hundred Steps*

There is a ladder that has 100 steps. One pigeon sits on the first step, two pigeons on the second, three pigeons on the third, four on the fourth, five on the fifth, and so on up to the hundredth step. How many pigeons are there in total on the ladder?

Solutions

13. *Answer*: 1,073,741,824

Solution

At the first village the servant starts by conscripting one man into the army. This can be represented with a power of 2, namely $2^0 (= 1)$. He leaves the village with 2^0 conscripts. Since he has conscripted one, he must pick up one more in the next village, for a total of 2. This can also be represented with a power of 2, namely 2^1. So, after two villages, he has picked up 2 conscripts. He then goes to the next village with the two conscripts, and picks up two more, for a total of four conscripts, which is 2^2. This is how many conscripts he has at this point. He enters the next village with the four conscripted men and picks up four more, for a total of 8, which is 2^3. And so on. In this way, he will have picked up 2^{30} or 1,073,741,824 conscripts in total.

42. *Answer*: 5050

Solution

Here is Alcuin's ingenious solution in the form of a breakdown:

- Add the 1 pigeon sitting on the first step to the 99 pigeons sitting on the 99th step:

$$1 + 99 = 100$$

- Add the 2 pigeons on the second step to the 98 pigeons on the 98th step.

$$2 + 98 = 100$$

- Add the 3 pigeons on the third step to the 97 pigeons on the 97th step.

$$3 + 97 = 100$$

- By adding the pigeons in this way, we will get 100 pigeons each time stopping at step 49.

$$\text{Step } 1 + \text{Step } 99 = 100$$
$$\text{Step } 2 + \text{Step } 98 = 100$$
$$\text{Step } 3 + \text{Step } 97 = 100$$
$$\dots$$
$$\text{Step } 49 + \text{Step } 51 = 100$$

- So far, the number of pigeons is:

$$49 \times 100 = 4900$$

- There are 50 pigeons on the 50th step, and 100 pigeons on the 100th step.
- Adding these to the 4900:

$$4900 + 50 + 100 = 5050 \text{ pigeons}$$

Annotations

Alcuin lived over several hundred years before Leonardo Fibonacci (1170–1250). But like Fibonacci, Alcuin was instrumental in making the concept of sequence known broadly with his two problems, as the *Propositiones* made their way into the hands of later mathematicians. The concept of sequences is an ancient one, found in manuscripts across the world, but its appearance in the form of these two problems in the *Propositiones* is remarkable on at least two counts—one of these is the coincidence between Problem 42 and Carl Friedrich Gauss's (1777–1885) solution to his schoolroom problem (discussed below); and another is that solving Problem 13 involves a series based on "2^n" as the general term, anticipating several other problems in recreational mathematics, including *Lucas's Towers of Hanoi Game* (also discussed below).

Gauss's Problem

The solution to Alcuin's Problem 42 is exactly like the solution to a famous one that is connected to German mathematician Carl Friedrich Gauss. Was the young Gauss aware of Alcuin's *Propositiones*? Or was it a serendipitous coincidence? It is impossible to connect the two problems historically, lacking textual evidence, but the parallels are notable. The well-known episode relates to Gauss as a nine-year-old child (Hayes 2009). In school one day, his teacher, a certain J. G. Büttner, asked the class to cast the sum of all the numbers from 1 to 100: $1 + 2 + 3 + 4 + \ldots + 100 = ?$ Gauss raised his hand within seconds, giving the correct response of 5050, astounding both his teacher and the other students who were toiling over the lugubrious arithmetical task before them. When Büttner asked Gauss how he was able to come up with the answer so quickly, he is said to have replied as follows:

> There are 49 pairs of numbers in the set of the first 100 numbers that add up to one hundred: $1 + 99 = 100, 2 + 98 = 100, 3 + 97 = 100$, and so on. That makes 4,900, of course. The number 50, being in the middle, stands alone, as does 100, being at the end. Adding 50 and 100 to 4,900 gives 5,050.

Impressed, the teacher not only arranged for Gauss's admittance to a school with a more challenging curriculum, but he also secured a tutor and advanced textbooks for the brilliant child. It is remarkable that Gauss's solution method is identical to Alcuin's for Problem 42—Gauss divided the numbers into "half sequences," from 1 to 49, and from 51 to 99, leaving 50 alone in the middle and 100 at the end, adding the first number in the first half (1) and last number in the second half (99), the second in the first half (2) and the second-last in the second half (98), and so on. This pattern, as we saw with Alcuin's problem, produces the constant sum of 100:

$$1 + 99 = 100$$
$$2 + 98 = 100$$
$$3 + 97 = 100$$
$$4 + 96 = 100$$
$$5 + 95 = 100$$
$$\ldots\ldots$$
$$49 + 51 = 100$$

The pairings adding up to 100 end at $49 + 51$. This means that there are 49 pairs that add up to 100. So, $49 \times 100 = 4900$. Adding to this the 50 and 100 that were isolated in the sequence makes $4900 + 50 + 100 = 5050$. The method used by both Alcuin and Gauss suggests a general rule for summing the series $\{1 + 2 + 3 + \ldots + n\}$, where n is any whole number. We start by representing the sum as:

$$S_1 = 1 + 2 + 3 + \ldots + (n - 2) + (n - 1) + n.$$

The same sum can be written with the numbers in reverse order:

$$S_2 = n + (n-1) + (n-2) + \ldots + 3 + 2 + 1$$

We can now line these up one under the other:

$$S_1 = 1 + 2 + 3 + \ldots + (n-1) + n$$
$$S_2 = n + (n-1) + (n-2) + \ldots + 3 + 2 + 1$$

By adding the terms in the order in which they occur in S_1 and S_2 we will get the expression $(n + 1)$ each time:

First terms in S_1 and S_2: $1 + n = (n + 1)$

Second terms in S_1 and S_2: $2 + (n-1) = (n + 1)$

Third terms in S_1 and S_2: $3 + (n-2) = (n + 1)$

...

Last terms in S_1 and S_2: $n + 1 = (n + 1)$

The result $(n + 1)$ occurs exactly n times, or $n(n + 1)$. Since $S_1 = S_2$, the process of adding the two series together is equivalent to $S_1 + S_2 = S + S = 2S$. Therefore:

$$2S = n(n + 1)$$
$$S = n(n + 1)/2$$

As this shows, the formula for summing such a series is $n(n + 1)/2$.

The Fibonacci Sequence

The *Fibonacci Sequence* is perhaps the most famous of all sequences inside and outside of mathematics. It derives from a problem about rabbits in a cage that Leonardo Fibonacci included in his 1202 book, *Liber Abaci*.

Box 8.1 Leonardo Fibonacci (c. 1170–1240)

Born in Pisa to a merchant, Fibonacci traveled with his father all over the Byzantine Empire in his youth, and he was sent by his father to study mathematics in the Empire. Later, he traveled throughout Egypt, Syria, and other places, learning about the decimal system used in some of the places he visited. In 1202, Fibonacci published a book titled *Liber Abaci* ("Book of the Abacus") to illustrate the practicality and efficiency of the decimal system to the European public. He was so taken by it that he would often write from right to left, in imitation of Semitic writing style, including the numerals in descending order {9, 8, 7, 6, 5, 4, 3, 2, 1, 0}.

The famous problem is found in the third section of the *Liber Abaci*. It is paraphrased below:

> A certain man put a pair of rabbits, male and female, in a very large cage. How many pairs of rabbits can be produced in that cage in a year if every month each pair produces a new pair which, from the second month of its existence on, also is productive?

The solution can be broken down as follows:

- A pair of rabbits is put in the cage at the start.
- *Number of pairs in cage at the start = 1*

- At the end of the first month, there is still only that one pair, because the problem states that a pair is productive only "from the second month of its existence on."
- *Number of pairs after one month = 1*

- It is during the second month that the original pair will produce its first offspring pair. Thus, at the end of the second month, a total of two pairs, the original one and the first offspring pair, are in the cage.
- *Number of pairs after second month = 2*

- During the third month, the original pair generates another new pair. The first offspring pair must wait a month before it becomes productive. So, at the end of the third month, there are three pairs in total in the cage—the initial pair, and the two offspring pairs that the original pair has thus far produced.
- *Number of pairs after third month = 3*

- At this point, let us step back and lay out the sequence of pairs that the cage successively contains from the start to the end of the third month as follows:

$$\{1, 1, 2, 3\}$$

- During the fourth month, the original pair produces yet another pair. At that point in time the first offspring pair produces its own offspring pair. The other pair has not started producing yet. Therefore, during that month, a total of two newborn pairs of rabbits are added to the cage. Altogether, at the end of the month there are five pairs in the cage: the previous three pairs plus the two newborn ones. This number can now be put into our sequence:

$$\{1, 1, 2, 3, 5\}$$

- During the fifth month, the original pair produces yet another newborn pair; the first offspring pair (now fully productive) produces another pair of its own as well (its second); and now the second offspring pair produces its first pair, for a total of three newborn pairs. The other rabbits in the cage have not started producing offspring yet. So, at the end of the month, three new pairs have been added to the five pairs that were previously in the cage, making the total number of pairs in it equal to eight. We can now put this number into our sequence:

$$\{1, 1, 2, 3, 5, 8\}$$

- Continuing to reason in this way, we will find that after 12 months, there are 233 pairs in the cage:

$$\{1, 1, 2, 3, 5, 8, 13, 21, 34, 55, 89, 144, 233\}$$

Examining the sequence closely, it can be seen that each number in it equals the sum of the previous two numbers: for example, 2 (the third number) = 1 + 1 (the sum of the previous two); 3 (the fourth number) = 1 + 2 (the sum of the previous two); etc. If we let F_n represent any "Fibonacci number" (any term in the sequence), F_{n-1} the number just before it, and F_{n-2} the number just before that, the pattern inherent in the sequence can be shown as follows:

$$F_n = F_{n-1} + F_{n-2}$$

This provides a "snapshot" of the internal structure of the sequence, which can be extended ad infinitum:

$$\{1, 1, 2, 3, 5, 8, 13, 21, 34, 55, 89, 144, 233, 377, 610, 987, \ldots\}$$

Little did Fibonacci know how significant his sequence would become. Over the years, the properties of the Fibonacci numbers have been studied extensively, resulting in a considerable literature. The rule generating the sequence was first studied by the French mathematician Albert Girard in 1632. In 1753, the Scottish mathematician Robert Simson noted that, as the terms increased in magnitude, the ratio of successive terms approached the golden ratio:

$$1/1 = 1.0$$
$$2/1 = 2.0$$
$$3/2 = 1.5$$
$$5/3 = 1.666...$$
$$8/5 = 1.6$$
$$13/8 = 1.625$$
$$21/13 = 1.6153...$$
$$34/21 = 1.6190...$$
$$55/34 = 1.6176...$$
$$89/55 = 1.6181...$$
$$......$$
$$F_{n-2}/F_{n-1} = 1.618...$$

Historians trace the discovery of the golden ratio to the Greek sculptor and mathematician Phidias (500–432 BCE), who applied it to the design of sculptures for the Parthenon. Euclid's *Elements* (Book VI) contains a formal definition of the ratio. The ratio was subsequently employed by Egyptian mathematician Abu Kamil (c. 850–930) to study polygons, likely influencing Luca Pacioli (c. 1447–1517) to study the sense of aesthetic beauty that seemingly results when this ratio is employed in architecture and art in his book *Divina proportione* (1509). Leonardo da Vinci illustrated Pacioli's book, calling it the *sectio aurea* (golden section).

Stretches of the Fibonacci sequence crop up in Nature. In most flowers, for example, the number of petals is one of the Fibonacci numbers 3, 5, 8, 13, 21, 34, 55, or 89—typically lilies have 3 petals, buttercups 5, delphiniums 8, marigolds 13, asters 21, daisies 34 or 55 or 89. In sunflowers, the little florets that become seeds in the head of the sunflower are arranged in two sets of spirals: one winding in a clockwise direction, the other counterclockwise. The Fibonacci number in the clockwise one is often 21 or 34 and in the counterclockwise one 34, 55, 89, or 144.

It was French mathematician François Edouard Anatole Lucas (1842–1891) who noticed the far-reaching significance of the sequence in the nineteenth century, coming up with similar sequences of his own in order to study their mathematical properties. In 1962, Verner Emil Hoggart and Brother Alfred Brousseau founded a Fibonacci Society for the purpose of studying the sequence and the patterns it concealed, as well as mathematical topics related to it. The Society started publication of a periodical, called the *Fibonacci Quarterly*, the year after in 1963. It continues to be published. As Morris Kline (1985: 42) aptly notes, not only are we "completely ignorant about the underlying reasons" for the manifestations of the Fibonacci sequence in many areas of reality and mathematics, but "we shall perhaps always remain ignorant of them."

Lucas's Towers of Hanoi Game

A famous game, invented by Lucas in 1883, is called the *Towers of Hanoi*. It is intriguing mathematically, given that, like Alcuin's Problem 13, the term, 2^n, shows up in the game. Below is Lucas's presentation of the game in the form of a conundrum:

> A monastery in Hanoi has a tower with three pegs. One holds 64 gold discs in descending order of size—the largest at the bottom, the smallest at the top. The monks have orders from God to move all the discs to the third peg while keeping them in descending order as they are moved, so that a larger disc must never sit on a smaller one. All three pegs can be used, and a disc can skip over a peg to the next one, and the moves can go in either direction. When the monks move the last disk, the world will end. Why?

The reason the "world will end" is that it would take the monks $(2^{64}-1)$ moves to accomplish the task, which is 18,446,744,073,709,551,615 moves. At one second per move, the universe would have disappeared long before the final move was made. Below is a do-able three-disc version of the game, in which three rectangular discs of increasing size are placed on the first of three pegs, with the smallest at the top, the mid-sized one just below, and the largest one on the bottom (Figure 8.1).

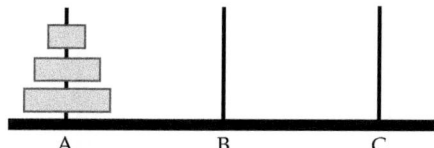

Figure 8.1 Three-Disc Version of the Towers of Hanoi Game here

The moves are as follows:

1. Move the smallest disc from A to C (Figure 8.2).

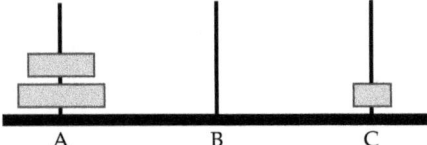

Figure 8.2 First Move

2. Move the mid-sized disc from A to B (Figure 8.3).

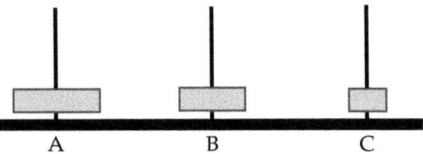

Figure 8.3 Second Move

3. Move the smallest disc on C, from C to B on top of the mid-sized one that is there (Figure 8.4).

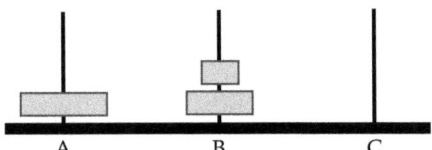

Figure 8.4 Third Move

4. Move the remaining disc on A (the largest one) from A to C (Figure 8.5).

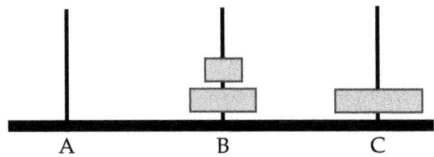

Figure 8.5 Fourth Move

5. Move the smallest disc on B from B back to A (Figure 8.6).

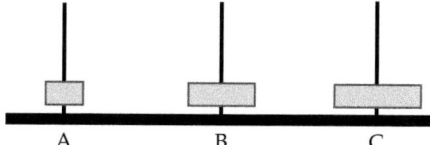

Figure 8.6 Fifth Move

6. Move the mid-sized disc remaining on B from B to C on top of the largest one there (Figure 8.7).

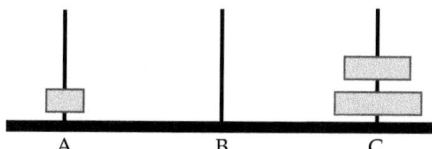

Figure 8.7 Sixth Move

7. Finally, move the smallest disc on A from A to C, on top of the discs there, completing the transfer of the discs from A to B according to the given rules (Figure 8.8).

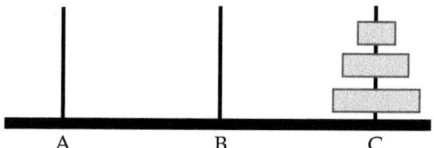

Figure 8.8 Seventh Move

It took seven moves to accomplish the task. The result can be represented as $(2^3-1) = 7$, noting that the exponent, "3," stands for the number of discs. If we have four discs to move instead of three, we will find that the number of moves is (2^4-1), indicated by the exponent of "4." If we were to play the game with any number of discs, "n," we would find that the number of moves is expressed by the general formula (2^n-1). In Lucas' original game, the number of discs is $n = 64$, so the number of moves needed to accomplish the task of transferring the discs from the first to the third peg would be $(2^{64}-1)$.

It is fascinating to note that the expression (2^n-1) surfaces in other areas of mathematics, going back to antiquity and its manifestation in the so-called *perfect* numbers. A perfect number is a number whose proper factors (all factors except the number itself), when added together, equal the number itself. For example, "6" is perfect because the numbers that divide into it are 1, 2, and 3, and when these are added together we get "6." Similarly, "28" is perfect because its proper factors are 1, 2, 4, 7, and 14, and when these are added together we get 28.

These numbers were studied by Euclid in his *Elements*, in which he showed that if $(2^n - 1)$ is prime, then $2^{(n-1)} (2^n - 1)$ is a perfect number—a truly remarkable demonstration, if one stops to think about it. Consider the first four perfect numbers—6, 28, 496, and 8128. Note that "n" is itself prime:

First perfect number: 6

$$n = 2$$
$$(2^n - 1) = (2^2 - 1) = (4 - 1) = 3 \text{ (prime number)}$$
$$2^{(n-1)} = (2^{2-1}) = 2^1 = 2$$
$$2^{(n-1)}(2^n - 1) = (2)(3) = 6$$

Second perfect number: 28

$$n = 3$$
$$(2^n - 1) = (2^3 - 1) = (8 - 1) = 7 \text{ (prime number)}$$
$$2^{(n-1)} = (2^{3-1}) = 2^2 = 4$$
$$2^{(n-1)}(2^n - 1) = (4)(7) = 28$$

Third perfect number: 496

$$n = 5$$
$$(2^n - 1) = (2^5 - 1) = (32 - 1) = 31 \text{ (prime number)}$$
$$2^{(n-1)} = (2^{5-1}) = 2^4 = 16$$
$$2^{(n-1)}(2^n - 1) = (16)(31) = 496$$

Fourth perfect number: 8128

$$n = 7$$
$$(2^n - 1) = (2^7 - 1) = (128 - 1) = 127 \text{ (prime number)}$$
$$2^{(n-1)} = (2^{7-1}) = 2^6 = 64$$
$$2^{(n-1)}(2^n - 1) = (64)(127) = 8128$$

Leonhard Euler proved centuries later that all even perfect numbers have this form. This is known as the Euclid–Euler theorem. It is unknown whether there are any odd perfect numbers, remaining one of the unsolved problems in number theory. In the seventeenth century, the French mathematician Marin Mersenne (1588–1648), used the expression (2^n-1) to generate a large number of primes, if "n" was itself prime. But it did not generate all primes. Here are a few cases:

$n = 2$ (prime)
$(2^n - 1) = (2^2 - 1) = 4 - 1 = 3$ (prime)

$n = 3$ (prime)
$(2^n - 1) = (2^3 - 1) = 8 - 1 = 7$ (prime)

$n = 5$ (prime)
$(2^n - 1) = (2^5 - 1) = 32 - 1 = 31$ (prime)

...

$n = 19$ (prime)
$(2^n - 1) = (2^{19} - 1) = 524,288 - 1 = 524,287$ (prime)

The prime numbers generated in this way are called *Mersenne primes*. By 1914, mathematicians had discovered a Mersenne prime that was 29 digits long. In 1996, computer programmer George Woltman started the Great Internet Mersenne Prime Search (GIMPS), with volunteers from across the globe searching for larger and larger Mersenne primes. The project continues to search for Mersenne primes and for perfect numbers as well.

The crystallization of the same expression in different areas of mathematics, from perfect numbers to a simple game with discs and pegs, is what makes mathematics truly intriguing. Maybe such discoveries are keys to understanding the structure of reality, or maybe they are constructs that we use to mirror the world and describe it on our own terms. Whatever the case, they would have remained unknown without puzzles and games such as those by Alcuin and Lucas, which fleshed them out into the open.

Infinite Sequences

As a final annotation, it is worth revisiting the notion of infinite sequence. A starting point is the paradoxes attributed to Zeno of Elea in the fifth century BCE. Recall from chapter 1 the *Achilles and the Tortoise* paradox, which was given as a problem in the *Explorations* section of the chapter.

In another paradox, called the *Dichotomy Paradox*, Zeno argued that if we use strict logic, then we must conclude that a runner will never reach the end of a race course, even though the runner will actually do so if we just look at the runner reaching the end. He argued this by stating that the runner first completes half of the course, then half of the remaining distance, and so on infinitely without ever reaching the end. If the length of the race course is represented by a line, with unit length, a few of the successive stages marking the runner's locations as he goes to the end line can be shown as follows (Figure 8.9).

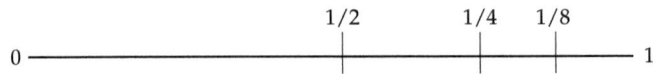

Figure 8.9 Stages of the Race Course

The successive stages form an infinite series with each term in it is half of the previous one.

$$\{1/2 + 1/4 + 1/8 + 1/16 + 1/32 + ...\} = 1/2^n$$

Zeno thus argued that the runner would seemingly never cross the end line, even though we know by experience that he actually would. Zeno's clever demonstration raised profound issues about time, space, and infinity. In it, we can easily discern the concept of *limits* which, in turn, inspired the invention of the calculus centuries later. Zeno's paradoxes treat distance and time as if they can be segmented into infinitely smaller parts or points.

With such shrewdly contrived arguments, Zeno threatened to tear down the whole edifice of Greek mathematics and philosophy, based on logic and argumentation. As Devlin (1998: 101) aptly puts it: "Zeno's puzzles presented challenges to the attempts of the day to provide analytic explanations of space, time, and motion—challenges that the Greeks themselves were not able to meet." Aristotle dismissed Zeno's paradoxes as exercises in specious reasoning. Contrary to Aristotle's critique, the paradoxes became important factors in the development of both mathematics and philosophy. As Kasner and Newman (1940: 39) perceptively stated, the "history of mathematics, in fact, recounts a poetic vindication of Zeno's stand."

Epilogue

Alcuin's ingenious Problems 13 and 42 show how a recreational approach to elementary mathematical notions can stimulate creative thinking. Moreover, by literally "playing around" with sequences, hidden patterns are often fleshed out of them, as was the case with Problem 42. The blueprint for Problem 42 may actually go back to Archimedes, as Hayes (2007) argues, since it is "not too much of a stretch to suppose that an educated monk in the court of Charlemagne might have some acquaintance with Archimedes."

The study of sequences has borne great fruit in mathematics. In his *Handbook of Integer Sequences* (1973), mathematician Neil Sloane provided the first database of 2372 integer sequences, in which he includes fascinating information about the sequences that may not be widely known. For instance, the sequence {1, 2, 4, 9, 21, 51, 127, . . . } was first studied in 1870, which represents the number of different ways of drawing non-intersecting chords between points on a circle (not necessarily touching every point by a chord). Over the years, Sloane (with collaborators) has come up with more and more distinct integer sequences, which he has uploaded to the Internet on the site, *Online Encyclopedia of Integer Sequences (OEIS)*. As of 2023, it contained more than 360,000 entries.

Explorations

The explorations involve problems dealing with sequences.

1. *Integer Sequence*

What is the next number in the following sequence?

$$\{1, 2, 4, 8, 16, 32, 64, \ldots\}$$

2. *Another Integer Sequence*

What is the next number in the following sequence?

$$\{1, 3, 9, 27, 81, 243, 729, \ldots\}$$

3. *A Unique Sequence*

What is the next number in the following sequence?

$$\{2, 3, 5, 7, 11, 13, 17, \ldots\}$$

4. *A Decimal Sequence*

What is the next number in the following sequence?

$$\{1, 0.5, 0.25, 0.125, 0.625, 0.03125, 0.015625, \ldots\}$$

5. *Sylvester's Sequence*

Can you state the rule for generating the following sequence?

$$\{2, 3, 7, 43, 1807, 3263443, ...\}$$

[This sequence was studied by the mathematician James Joseph Sylvester in 1880.]

6. *Fibonacci Redux*

Recall the Fibonacci sequence:

$$\{1, 1, 2, 3, 5, 8, 13, 21, 34, 55, ...\}$$

Here is a sequence derived from it:

$$\{1, 1, 4, 9, 25, 64, 169, 441, 1156, 3025, ...\}$$

What is the numerical pattern in the new sequence?

7. *Embedded Sequence*

What is the next term in the following sequence?

$$\{2, 3, 4, 9, 8, 27, 16, 81, 32, 243, 64, 729...\}$$

8. *A Tricky Sequence*

What is the next term in the following sequence?

$$\{2, 3, 5, 9, 17, 33, 65, ...\}$$

9. *Another Tricky Sequence*

Can you figure out what the next term in the following sequence is?

$$\{0, 1, 3, 7, 15, 31, 63, ...\}$$

10. *The Lucas Sequence*

Édouard Lucas, as mentioned, created his own sequences (Vajda 1989). In one of them he started with the terms 2 and 1:

$$\{2, 1, 3, 4, 7, 11, 18, 29, 47, 76, 123, ...\}$$

Can you figure out the rule that generates each term?

Cited Works and Further Reading

Basin, S. L. (1963). The Fibonacci Sequence as It Appears in Nature. *The Fibonacci Quarterly* 1: 53–64.

Bindner, Donald J. and Erickson, Martin (2012). Alcuin's Sequence. *The American Mathematical Monthly* 119: 115–121.

Brousseau, Brother A. (1965). *An Introduction to Fibonacci Discovery*. Aurora, SD: The Fibonacci Association.

Burkholder, Peter (1993). Alcuin of York's *Propositiones ad acuendos juvenes*: Introduction, Commentary & Translation. *History of Science & Technology Bulletin*, Vol. 1, number 2.

Devlin, Keith (1998). *The Language of Mathematics: Making the Invisible Visible*. New York: W. H. Freeman.

Devlin, Keith (2011). *The Man of Numbers: Fibonacci's Arithmetic Revolution*. New York: Walker and Company.

Dunlap, Richard A. (1997). *The Golden Ratio and Fibonacci Numbers*. Singapore: World Scientific.

Fibonacci, Leonardo (1202). *Liber Abaci*, trans. by L. E. Sigler. New York: Springer, 2002.

Friedmann, Tamar and Hagen, Carl R. (2015). Quantum Mechanical Derivation of the Wallis Formula for π. *Journal of Mathematical Physics* 56: https://doi.org/10.1063/1.4930800.

Gardner, Martin (1969). The Multiple Fascination of the Fibonacci Sequence. *Scientific American*, March (1969), 116–120.

Garland, Trudi H. (1987). *Fascinating Fibonaccis*. White Plains, NY: Dale Seymour Publications.

Hayes, Brian (2006). A Famous Story About the Boy Wonder of Mathematics Has Taken on a Life of Its Own. *American Scientist*, americanscientist.org/article/gauss-day-of-reckoning.

Hayes, Brian (2007). A Mathematical Fable Previsited, http://bit-player.org/2007/a-mathematical-fable-revisited.

Jean, Roger V. (1984). *Mathematical Approach to Pattern in Plant Growth*. New York: John Wiley.

Kasner, Edward and Newman, James (1940). *Mathematics and the Imagination*. New York: Simon and Schuster.

Livio, Mario (2002). *The Golden Ratio: The Story of Phi, the World's Most Astonishing Number*. New York: Broadway Books.

Lucas, Édouard (1892). *Récréations mathématiques*. Paris: Albert Blanchard.

Maor, Eli (1987). *To Infinity and Beyond: A Cultural History of the Infinite*. Boston: Birkhäuser.

O'Connor, John J. and Robertson, Edmund F. (2012). *Propositiones ad acuendos juvenes*. https://mathshistory.st-andrews.ac.uk/HistTopics/Alcuin_book/.

Sloane, Neil (1973). *Handbook of Integer Sequences*. New York: Academic Press.

Stewart, Ian (1987). *From Here to Infinity: A Guide to Today's Mathematics*. Oxford: Oxford University Press.

Vajda, Steven (1989). *Fibonacci and Lucas Numbers, and the Golden Section*. Chichester: Ellis Horwood.

Varnadore, James (1991). Pascal's Triangle and Fibonacci Numbers. *The Mathematics Teacher* 84. 314–316.

9

Measurement

Prologue

The earliest units of weights and measures are traced to the third millennium BCE. They were used for practical purposes in agriculture, building, trade, and science. As trade between the early civilizations became increasingly important, standardizing the units became critical. By the eighteenth century, uniform systems of measurement were finally established. The most common one, especially in science, is the metric system, which uses the meter, liter, and gram as base units of length (distance), capacity (volume), and weight (mass), respectively. The system is also called "decimal" because it is based on multiples of ten. Any measurement given in one metric unit can be converted to another metric unit simply by a multiple of ten. For example, a kilogram equals 1000 grams, and a millimeter 1/1000 of a meter.

Using mathematics to generalize practical measurement problems has always been a central part of education since antiquity, as the many measurement problems in the *Rhind Papyrus* and the *Greek Anthology* confirm. Some of these appear to feed into Alcuin's own problems, suggesting that they had become commonplace across ancient and medieval societies. However, there is one problem in Alcuin's set—involving pouring from one container to another—that appears to have its first appearance in the *Propositiones* (Cowley 1926).

The Problems

Seven of the problems in the *Propositiones* involve measurement—numbers 8, 9, 10, 12, 30, 31, and 51.

8. *Problem of the Cracked Cask*

There is a cask with three cracks in it. It is filled with 7200 pints of water. A third plus a sixth part of the water runs out through one crack. Through another crack a third part runs out. And a sixth part runs out through the third crack. How many pints ran out through each crack?

9. *Problem of the Cloaks*

I have material that is 100 feet long and 80 feet wide. From it, I wish to make cloaks in such a way that each cloak is five feet in length and four feet in width. How many such cloaks can I make from the material?

10. *Problem of the Tunics*

I have a single linen cloth which is 60 feet long and 40 feet wide. I wish to cut it into smaller pieces to make tunics, each one six feet in length and four feet in width. How many such tunics can I make from the linen cloth?

12. *Problem of the Father and His Three Sons*

A father wants to leave, as an inheritance to his three sons, 30 glass flasks: 10 are full of oil, another 10 are half full, while the last 10 are empty. How should he divide the flasks so that an equal share of oil will come down to the three sons?

30. *Problem of the Basilica*

A basilica is 240 feet long and 120 feet wide. It is to be paved with tiles measuring 1 foot, 11 inches long by 1 foot wide. How many tiles are needed?

31. *Problem of the Wine Cellar*

A wine cellar is 100 feet long and 64 feet wide. How many casks can it hold, given that each cask is seven feet long and four feet wide, and given that there is an aisle four feet wide down the middle of the cellar?

51. *Problem of the Inherited Flasks*

A father wants to leave four flasks of wine to divide among his four sons. In the first flask, there are 40 measures of wine, in the second there are 30 measures of wine, in the third there are 20 measures of wine, and in the fourth there are 10 measures of wine. Calling his house treasurer, the father instructed him: "Divide these four flasks amongst my four sons in such a way that each son receives an equal portion of wine and flasks." The servant had no means of measuring wine and no container other than the four flasks. How did he carry out the dying father's wishes?

   [To solve the problem, assume that the flasks are calibrated in units that can be seen on them.]

Solutions

8. *Answer:* first crack, 3600 pints; second crack, 2400 pints; third crack, 1200 pints

Solution

Here is the breakdown.

- *First crack:* $1/3 + 1/6 = 1/2$ of the water runs through the first crack. This means that $1/2 \times 7200 = 3600$ pints flowed out through that crack.
- *Second crack:* $1/3 \times 7200 = 2400$ pints flowed out of the second crack.
- *Third crack:* $1/6 \times 7200 = 1200$ pints ran out through the third crack.

9. *Answer:* 400

Solution

Here is the breakdown.

- The material is $100 \times 80 = 8000$ (square feet).
- Each cloak is to be $5 \times 4 = 20$ (square feet).
- Thus, the number of cloaks that can be made is: $8000 \div 20 = 400$.

10. *Answer:* 100

Solution

Here is the breakdown.

- The linen cloth is $60 \times 40 = 2400$ (square feet).
- Each tunic is to be $6 \times 4 = 24$ (square feet).
- So, the number of tunics that can be made is: $2400 \div 24 = 100$.

12. *Answer:* see solution

Solution

The 30 flasks can be divided equitably as follows:

First son:
10 half flasks = 5 full flasks
Second son:
5 full and 5 empty flasks = 5 full flasks
Third son:
5 full and 5 empty flasks = 5 full flasks

This is a truly interesting problem, whose generalization has come to be known as *Alcuin's Sequence*, discussed below (Binder and Erikson 2012).

30. *Answer:* 15,120 (Alcuin's answer); 15,026 (answer based on measurements given)

Solution

Here is the breakdown.

- The basilica is 240 feet long and 120 feet wide. In inches, it is 240 × 12 = 2880 inches long by 120 × 12 = 1440 inches wide.
- So, its area is: 2880 × 1440 = 4,147,200 (square inches).
- It is to be paved with tiles measuring 1 foot, 11 inches long by 1 foot wide. In inches, each tile is 23 inches by 12 inches.
- The area of each tile is: 23 × 12 = 276 (square inches).
- So, the number of tiles required is: 4,147,200 ÷ 276 = 15,026.

31. *Answer:* 210 (Alcuin's answer); 200 (possible); 214 (possible)

Solution

Here is one possible breakdown.

- Without the aisle, the cellar would be 100 × 60 = 6000 (square feet).
- The aisle goes through the cellar, and is thus 100 feet long and 4 feet wide (as given). It thus covers 100 × 4 = 400 (square feet).
- Taking away the aisle, this leaves 6000 − 400 = 5600 (square feet).
- Each cask is 7 × 4 = 28 (square feet).
- So, the number of casks that the cellar can hold is: 5600 ÷ 28 = 200.
- Without going into details here, Alcuin gives the answer of 210, which he reaches by laying tiles along the two equal rectangular strips of the cellar that remain after the aisle is removed.
- There are other ways to solve the problem. For example, O'Connor and Robertson (2012) came up with 214. All this shows that the problem was much more "subtle" than what Alcuin may have imagined, as O'Connor and Robertson observed.

51. *Answer:* see solution

Solution

Here is the breakdown.

- The total amount of wine is 40 + 30 + 20 + 10 = 100 measures.
- Since each son is to receive an equal portion of wine, this means that each one must receive 25 measures of wine.
- Now, the problem for the treasurer is to devise a way of getting 25 measures into each flask without being able to measure it.
- The flasks are all the same dimensions. So, here is how he does it, tipping wine from one to the other until each has an equal amount of wine in each. Assume that each flask is calibrated in unit measures that are inscribed on it. Each flask is full at 40 measures.

   1. He pours 30 measures from the 40-measure flask into the 10-measure flask. This means that 10 measures are left in

the 40-measure flask, while the 10-measure flask now has 40 measures.

2. He then pours 20 measures from the 40 in the 10-measure flask back into the 40-measure one. The end result is that the 40-measure flask now has $10 + 20 = 30$ measures and the 10-measure flask has $40 - 20 = 20$ measures.

3. Finally, he pours 5 measures from the 40-measure flask into the 10-measure one. This means that the 40-measure flask is left with $30 - 5 = 25$ measures and the 10-measure one with $20 + 5 = 25$ measures.

- The exchanges between these two flasks are summarized below:

40 – Measure Flask		10 – Measure Flask
$40 - 30 = 10$	$30\rightarrow$	$10 + 30 = 40$
$10 + 20 = 30$	$\leftarrow 20$	$40 - 20 = 20$
$30 - 5 = 25$	$5\rightarrow$	$20 + 5 = 25$

- These now contain 25 measures each.
- The treasurer now takes the flasks containing 30 measures and 20 measures and tips wine from one into the other. Only one pouring is required this time—5 measures from the 30-measure flask are poured into the 20-measure flask, meaning that 25 measures are left in each:

30 – Measure Flask		20 – Measure Flask
$30 - 5 = 25$	$5\rightarrow$	$20 + 5 = 25$

- All the flasks now contain 25 measures. Note that this solution assumed unit calibrations on each flask. Other solutions do not.

Annotations

The inclusion of measurement problems in the *Propositiones* was consistent with traditions in mathematics pedagogy since antiquity. They present various interesting features, though, as the solutions bring out, that are exceptional for medieval math texts. The liquid-pouring problem (number 51) is unique, since there appear to be no precedents to it before Alcuin. It became a staple of recreational mathematics thereafter (Tweedie 1939). One version was Bachet's weighing puzzle, which was given in this book as an exploration in Chapter 2; another one was by Niccolò Tartaglia, which will be given as an exploration at the end of this chapter. Two other famous descendants of Alcuin's problem were by Nicolas Chuquet and Lewis Carroll, discussed below.

Chuquet's Version

The fifteenth-century mathematician Nicolas Chuquet (1445–1488) included a transfer puzzle in his 1484 book, *Triad on the Science of Numbers*. Below is a paraphrase.

> You have two empty casks which can hold 5 and 3 pints respectively. There is also a barrel that has a huge amount of liquid in it that can be used at any time. How can you measure exactly 4 pints given that you are allowed to pour liquid back and forth between the casks?

Here is a breakdown:

1. Fill the 5-pint cask from the barrel (Figure 9.1).

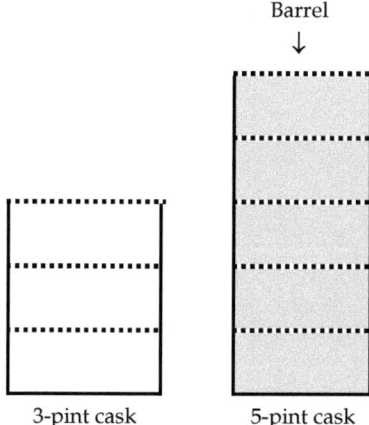

3-pint cask   5-pint cask

Figure 9.1 First Transfer

2. Pour 3 pints from the full 5-pint cask into the 3-pint one. This fills the 3-pint cask and leaves 2 pints in the 5-pint one (Figure 9.2).

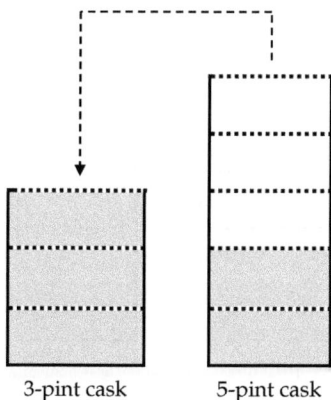

3-pint cask   5-pint cask

Figure 9.2 Second Transfer

3. Empty the 3-pint cask (Figure 9.3).

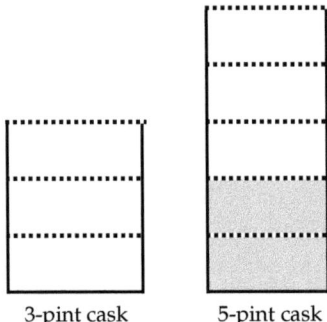

3-pint cask   5-pint cask

Figure 9.3 Third Transfer

4. Pour the 2 pints that are in the 5-pint cask into the 3-pint cask. This leaves the 5-pint cask empty (Figure 9.4).

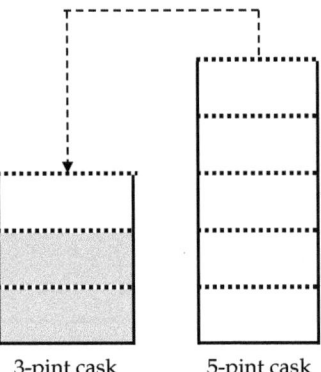

3-pint cask   5-pint cask

Figure 9.4 Fourth Transfer

5. Fill the 5-pint cask from the barrel (Figure 9.5).

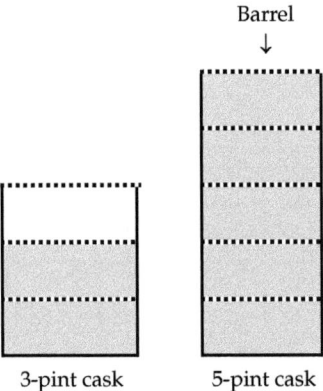

3-pint cask   5-pint cask

Figure 9.5 Fifth Transfer

6. Pour liquid into the 3-pint cask from the 5-pint cask so as to fill it. This will add a pint to the 3-pint cask and leave 4 pints in the 5-pint cask, which is the required solution (Figure 9.6).

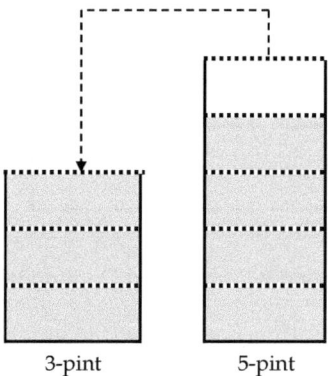

3-pint   5-pint

Figure 9.6 Sixth Transfer

Such puzzles involve a finite collection of containers of known integer capacities. Initially, each one contains a known integer volume of liquid, not necessarily equal to its capacity, or no liquid at all. These then ask how many steps of pouring liquid from one container to another are needed to reach a goal state, specified in terms of the volume of liquid that must be present in some containers. For example, three containers of capacity 8, 5, and 3 liters are initially filled with 8, 0, and 0 liters. The goal state might be to fill them with 4, 4, and 0 liters, respectively. This puzzle is solved in seven steps, passing through the following sequence of states (shown as bracketed triples of the three volumes of liquid):

$$[8,0,0] \rightarrow [3,5,0] \rightarrow [3,2,3] \rightarrow [6,2,0] \rightarrow [6,0,2] \rightarrow [1,5,2] \rightarrow [1,4,3] \rightarrow [4,4,0]$$

Carroll's Version

Lewis Carroll's three-container version is one of the best-known puzzles in recreational mathematics (Carroll 1958, Ball 1896, Singmaster 2005).

There are two containers on a table, labeled A and B. Container A is half full with wine, while B, which is twice A's size, is one-quarter full with wine. Both containers are filled with water and the contents are poured into a third container, C. What portion of container C's mixture is wine?

A sketch of the two containers, empty, would show that B is twice the size of A. Note that A is marked with two equal parts and B with four equal parts (Figure 9.7).

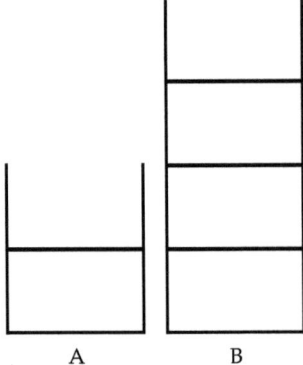

Figure 9.7 The Two Containers

Container A is half full with wine, while B is one-quarter full with wine. We note that this is the same amount of wine (Figure 9.8):

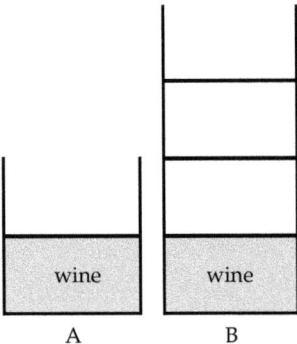

Figure 9.8 Level of Wine in the Two Containers

Next, we fill the containers with water (Figure 9.9).

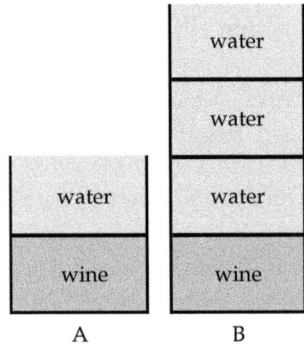

Figure 9.9 Containers with Wine and Water

As can be seen, A has two equal portions of wine and water, while B has three parts water and one part wine. Between the two containers, there are six equal parts in total—two parts wine and four parts water. A mixture of these two containers will contain two parts wine and four parts water. That is, in fact, what container C will have when the contents of A and B are poured into it (Figure 9.10).

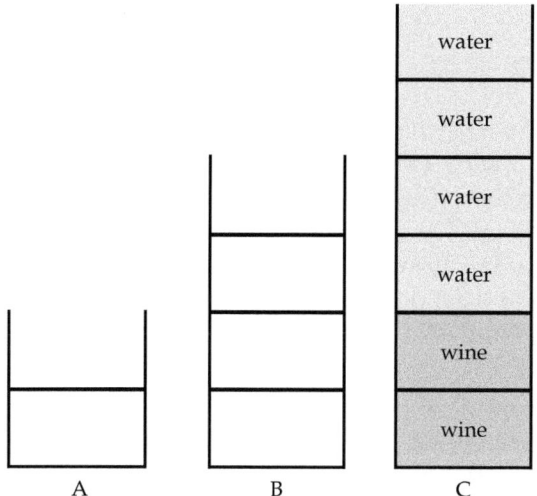

Figure 9.10 Transfer to Container C

The wine and water in container C will, of course, be mixed up, not separated neatly like we have shown in the diagram above. But in that mixture, wine will make up two parts out of its six, or 2/6 (= 1/3); and water will make up four parts out of its six, or 4/6 (= 2/3). In conclusion, C's mixture will have 2/6 = 1/3 wine in it.

Alcuin's Sequence

The general implications of Alcuin's Problem 12 have been examined by various mathematicians, in terms of its sequence structure, known as *Alcuin's Sequence*—a term first used by Dominic Olivastro in his book *Ancient Puzzles* (1993). The sequence is represented as follows (Weinstein 2004, Bindner and Erickson 2012):

$$1/(1-n^2)(1-n^3)(1-n^4) = 1 + n^2 + n^3 + 2n^4 + n^5 + 3n^6 + \ldots$$

Without going into the technical details here, which are beyond the present purposes, suffice it to say that it expresses the number of ways in which "*n*" empty casks, "*n*" casks half-full, and "*n*" full casks can be distributed in such a way that each of three persons gets the same number of casks and the same amount of liquid. There have been other implications derived from Alcuin's Sequence (Andrews 1979). Again, as in other areas of mathematics, this shows how a simple puzzle can lead to important mathematical insights and, in some cases, new branches of the discipline.

Epilogue

Mathematical ideas often come from measurement activities that harbor patterns within them—extracting the patterns formally is at the core of mathematical method. The early mathematicians refined the practical facts of measurement into geometrical-numerical models, allowing them to describe the world in abstract ways and thus to measure things that went beyond the world itself. Ian Stewart (2008: 46) summarizes this amazing achievement as follows:

> Using geometry as a tool, the Greeks understood the size and shape of our planet, its relation to the Sun and the Moon, even the complex motions of the remainder of the solar system. They used geometry to dig long tunnels from both ends, meeting in the middle, which cut construction time in half. They built gigantic and powerful machines, based on simple principles like the law of the lever.

Explorations

The problems in this section involve figuring out measurements, admixtures, or carrying out transfers from one container to another. Some of these are staples of recreational mathematics.

1. *Plotting a Field*

A farmer has a field measuring 50 by 40 feet. He plows two furrows from the top of the longest side to the other end, each 5 feet wide. He wants to lay down square plots of grass in the remaining field area. Each plot is 4 feet long by 4 feet wide. How many plots can be planted in the field?

2. *Cups of Liquid*

How many cups does my friend have if all of them are filled with soda except two, all are filled with water except two, and all are filled with milk except two?

3. *A Mixture Problem*

A farmer took a bunch of his potatoes and beets to a city market-place. The city government had put a balance scale there that merchants could use for their transactions. The farmer wanted to sell his potatoes at 45 cents per pound, and his beets at 30 cents per pound. A customer bought a mixture of 50 pounds of the two. The mixture sold for 40 cents a pound. How many pounds of each were there in the mixture?

4. *Another Mixture Problem*

A mixture of 12 ounces of vinegar and oil consists of 40% vinegar (by weight). How many ounces of oil must be added to that mixture to produce a new mixture with only 25% vinegar in it?

5. *Another Field Problem*

A farmer has a rectangular field, 100 feet long by 80 feet wide, in which he wants to plant a number of trees. However, he wants to section out a square area within the field, measuring 50 feet per side, to keep for grazing. He will put the trees in the remaining area. Each tree should occupy a 5-by-5 plot in that area. How many trees can he plant?

6. *Overflow Problem*

There are two faucets, a cold and a hot one, filling a tub, which can hold 520 liters of water. The cold water flows at the rate of 15 liters per minute, and the hot one at 10 liters per minute. A hole in the tub lets out water at 12 liters per minute. With the two faucets on at the same time, how long will it be before the bathtub begins to overflow?

7. *A Water Transfer Problem*

There are three jars, with capacity of 8, 5, and 3 liters each. Let us name them A, B, and C, respectively. Pouring water back and forth among the jars, make it so that the A-jar ends up with 3 liters in it and the other two with 5 and 0 liters in a minimum number of water transfers. Only the 8-liter jar is initially filled with water, while the other two are empty. What is the minimum number of transfers required?

[This problem is a paraphrase of an original problem by Niccolò Tartaglia, discussed by Dudeney (1917).]

8. *Another Water Transfer Problem*

How can 6 liters of water be measured exactly using only 4-liter and 9-liter jugs, which are empty at the start?

9. *One Final Transfer Problem*

This time how can you measure exactly 2 liters of water if you have empty 4-liter and 3-liter jugs in the least number of moves?

10. *A Faucets Problem*

A swimming pool has four faucets that fill it with water. The first faucet takes two days to fill the pool, the second three days, the third four days, and the fourth 6 hours. How long will it take to fill the pool using all faucets at once?

Cited Works and Further Reading

Andrews, George E. (1979). A Note on Partitions and Triangles with Integer Sides. *American Mathematical Monthly* 86: 477.

Ball, W. W. Rouse (1896). *Mathematical Recreations and Essays*, 12th edition, revised by H. S. M. Coxeter. Toronto: University of Toronto Press, 1972.

Beauregard, Raymond A. and Dobrushkin, Vladimir A. (2013). Finite Sums of Alcuin Numbers. *Mathematics Magazine* 86: 280–287.

Bindner, Donald J. and Erickson, Martin (2012). Alcuin's Sequence. *The American Mathematical Monthly* 119: 115–121.

Burkholder, Peter (1993). Alcuin of York's *Propositiones ad acuendos juvenes*: Introduction, Commentary & Translation. *History of Science & Technology Bulletin* 1(2).

Carroll, Lewis (1958). *Mathematical Recreations of Lewis Carroll*. New York: Dover.

Chuquet, Nicolas (1484). *Triparty en la science des nombres*. Paris: Aristide Marre.

Cowley, Elizabeth B. (1926). Note on a Linear Diophantine Equation. *American Mathematical Monthly* 33: 379–381.

Dudeney, Henry E. (1917). *Amusements in Mathematics*. London: Thomas Nelson and Sons.

O'Connor, J. J. and Robertson, E. F. (2012). *Propositiones ad acuendos juvenes*. https://mathshistory.st-andrews.ac.uk/HistTopics/Alcuin_book/.

Olivastro, Dominic (1993). *Ancient Puzzles: Classic Brainteasers and Other Timeless Mathematical Games of the Last 10 Centuries*. New York: Bantam.

Singmaster, David (2005). Mathematical Recreations and Problems of Past and Present Times. In: Ivor Grattan-Guinness and Roger Cooke (eds.), *Landmark Writings in Western Mathematics 1640–1940*. Oxford: Elsevier.

Stewart, Ian (2008). *Taming the Infinite*. London: Quercus.

Tweedie, M. C. K. (1939). A Graphical Method of Solving Tartaglian Measuring Puzzles. *The Mathematical Gazette* 23: 278–282.

Weinstein, Eric W. (2004). Alcuin's Sequence. *MathWorld*. https://mathworld.wolfram.com/AlcuinsSequence.html.

Woolhouse, Wesley Stoker Barker (1979). *Historical, Measures, Weights, Calendars & Moneys of All Nations and an Analysis of the Christian, Hebrew and Muhammadan Calendars*. Chicago: Ares Publishers.

10
Assembly and Partitioning Problems

Prologue

Several of the problems discussed in previous chapters involved fitting figures, such as smaller squares, into larger figures, such as larger squares, or partitioning figures in specific ways. This chapter deals with three similar problems, but which bear several important theoretical implications for mathematics generally. Such problems seem to have been of great interest in the medieval era, when cities were being designed according to different geometrical shapes, leading to mathematical investigations of the architectural styles of the cities and of the houses that were built within them.

Alcuin was likely aiming to impress upon his readers that three primary geometrical figures—a quadrilateral, a triangle, and a circle—constituted important figures in architecture and city design. Each type of shape can contain recurring structures within it, as the three problems in this chapter bring out. These involve situations that are based on some optimal design. In a relevant article, Nikolai Yu Zolotykh (2013) actually considers these problems to be the earliest examples of abstract packing problems in mathematics, given that they involve the placement of figures according to an optimal pattern.

The Problems

The relevant three problems in the *Propositiones* are numbers 27, 28, and 29.

27. *Problem of the Quadrangular City*

There is a four-sided city with one side measuring 1100 feet, another side 1000 feet, a front 600 feet, and a final side 600 feet. I want to put some houses there so that each house is 40 feet long and 30 feet wide. How many houses ought the city to contain?

28. *Problem of the Triangular City*

There is a triangular city that has one side of 100 feet, another side of 100 feet, and a third side of 90 feet. Inside this city, I want to build houses each of which is 20 feet in length and 10 feet in width. How many houses can I build in the city?

29. *Problem of the Round City*

There is a city which is 8000 feet in circumference. How many houses could the city contain if each house is to be 30 feet long and 20 feet wide?

Solutions

27. *Answer*: 520 (Alcuin's answer); 523 (more likely answer)

Solution

It is hard to envision what Alcuin had in mind geometrically, given the kind of confusing description he used. O'Connor and Robertson (2012) assume that the city is an isosceles trapezium, as in Figure 10.1.

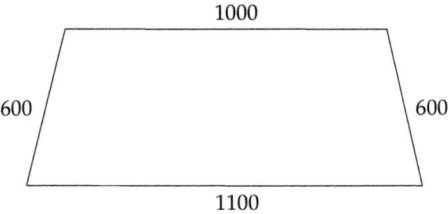

Figure 10.1 Possible Form of Alcuin's City

The area of a trapezium is: $(a + b)/2 \times h$, with a and b being the two parallel sides and h the height. In Alcuin's city, the height, h, can be figured out as follows:

- If we drop two perpendiculars, we will produce two right-angle triangles that dissect the base as shown in the diagram (Figure 10.2).

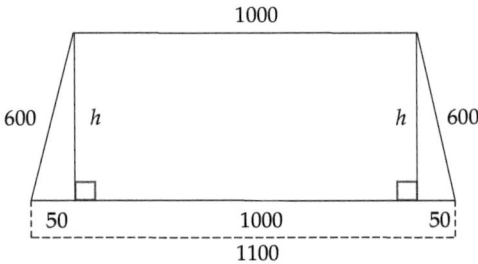

Figure 10.2 Triangles Constructed on Trapezium

- This is so because the perpendiculars are equal in length (between two parallel lines), and they are also sides of two congruent triangles, with dimensions h and 600.

- This means that the third side of each triangle is equal and thus 50 units in size, since the base of the trapezium is 1100, as shown in Figure 10.2.
- We can now use the Pythagorean theorem to determine the height:

$$600^2 = 50^2 + h^2$$
$$h^2 = 600^2 - 50^2$$
$$h^2 = 360,000 - 2500$$
$$h^2 = 357,500$$
$$h = 598 \text{ square feet (rounded off)}$$

- We can now compute the area of the trapezium:

$$\text{Area} = (a + b)/2 \times h$$
$$a = 1000$$
$$b = 1100$$
$$h = 598$$
$$\text{Area} = (1000 + 1100)/2 \times 598$$
$$\text{Area} = (2100)/2 \times 598$$
$$\text{Area} = 627,900 \text{ square feet}$$

- Each house will cover an area of $40 \times 30 = 1200$ square feet.
- Thus, the number of houses that can be made to fit into the city is $627,900 \div 1200 = 523$ (rounded off).
- Alcuin comes up with an answer of 520 houses. But it is hard to figure out how he envisioned the city in the first place.

28. *Answer*: 20

Solution

The dimensions of the city form an isosceles triangle, ABC, with two sides measuring 100 feet each, and the third side 90 feet (Figure 10.3).

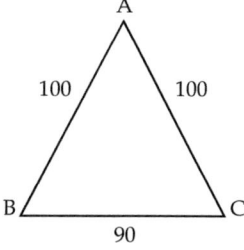

Figure 10.3 Outline of Alcuin's Triangular City

- We drop a perpendicular, h, from A to the base BC. Since it is an isosceles triangle the perpendicular will bisect that base (Figure 10.4).

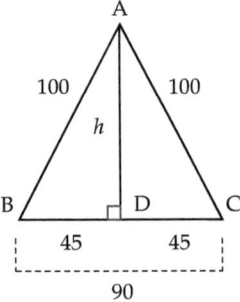

Figure 10.4 Perpendicular Dropped on Triangle ABC

- The area of a triangle is: ½ b (base) $\times h$ (height).
- So, we need to figure out first the height (perpendicular line AD) of the triangle, using the Pythagorean theorem:

$$AB^2 = BD^2 + h^2$$
$$100^2 = 45^2 + h^2$$
$$h^2 = 100^2 - 45^2$$
$$h^2 = 10,000 - 2025$$
$$h^2 = 7975$$
$$h = 89 \text{ square feet (rounded off)}$$

- So, the area of the triangular city is:

$$\text{Area} = {}^1\!/_2\, b \times h$$
$$b = 90$$
$$h = 89$$
$$\text{Area} = {}^1\!/_2 \times 90 \times 89 = 4005 \text{ square feet}$$

- Each house covers $20 \times 10 = 200$ square feet.
- So, the number of such houses that can be made to fit into the city is $4005 \div 200 = 20$ (rounded off).

29. *Answer*: 6400 (Alcuin's answer), 8498 (more likely)

Solution

The city is a circular one. The breakdown is as follows:

- First, we determine the length of the radius, r, of the circular city, given that the circumference (C) is 8000 feet:

$$C = 2\pi r$$
$$8000 = 2\pi r$$
$$r = 8000/2\pi$$
$$r = 1274 \text{ feet (rounded off)}$$

- So, the area of the circular city is:

$$A = \pi r^2$$
$$A = \pi \times 1274^2$$
$$A = 5,099,043 \text{ square feet (rounded off)}$$

- Each house covers $30 \times 20 = 600$ square feet.
- So, the number of houses that can be made to fit into the city (mostly) is $5,099,043 \div 600 = 8498$.
- This contrasts with Alcuin's answer of 6400. Again, it is hard to imagine what Alcuin actually had in mind.

Annotations

As inconsequential as they may seem on the surface, these three problems are governed by an important mathematical question: How can a certain geometric figure be tessellated internally with other figures systematically? As Zolotykh (2013) points out, these are essentially packing problems in contemporary mathematical terms.

In recreational mathematics, these may have been the inspiration for a host of subsequent puzzles dealing with assembling figures, as Zolotykh indirectly suggests. However, the prototype for this kind of puzzle, as discussed in Chapter 3, is Archimedes' loculus (Darling 2004, Netz and Noel 2007). There are many parallels between the loculus and modern-day jigsaw and tangram puzzles. The cut-out shapes could also be used to form different figures (human shapes, animals, objects, etc.)—an objective that applies as well to tangrams (discussed below).

Assembly Puzzles

Assembly puzzles and games have, since at least Alcuin, become staples of recreational mathematics. Consider a challenging game called *polyominoes*, which was invented by Solomon W. Golomb (1965), consisting of the shapes formed by connecting a number of unit squares edge-to-edge.

For example, the *domino* is a two-square figure, the *tromino* a three-square figure, the *tetromino*, a four-square figure, the *pentomino* a five-square figure, and so on. Below are pentomino shapes (Figure 10.5).

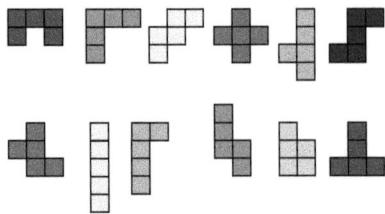

Figure 10.5 Pentominoes

Here is an example of a puzzle using these pieces:

Can pentominoes be assembled in such a way to produce an 8-by-8 square, with a 2-by-2 opening in the center of the square, using each piece once and only once?

It can be done as shown in Figure 10.6.

Figure 10.6 Solution to the Pentomino Puzzle

One of the best-known of all assembly puzzles in recreational mathematics is the *Tangram*. Some attribute its origin to ancient China (Vorderman 1996, Slocum, Botermans, et al. 2004); while others believe that it is of Japanese origin because of its appearance in a remarkable little book containing seven-piece tangram puzzles, titled *The Ingenious Pieces of Sei Shonagon*, that was published in 1742 in Japan (Takagi 1999). Whatever the truth, the Tangram made its way to Europe and America in the early nineteenth century, where it became extremely popular. It is said that even Napoleon was a tangram enthusiast while in exile on St. Helena.

There are seven tangram pieces—five triangles, one square, and one parallelogram inserted in a square (Figure 10.7).

Figure 10.7 The Tangram Puzzle

The objective is to assemble these pieces to produce recognizable shapes, figures, and forms. The seven pieces may be used as many times as needed. The challenge lies in assembling them to make patterns or figures. Here are two examples:

Make a horse figure with tangram pieces (from Loyd 1952) (Figure 10.8).

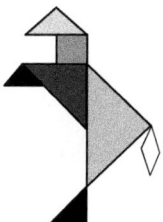

Figure 10.8 A Horse Figure

Now, make a house shape (Figure 10.9).

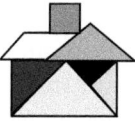

Figure 10.9 A House Figure

Packing Problems

As mentioned, Zolotykh (2013) interprets Alcuin's problems in terms of current-day packing problems, or optimization theory. Zolotykh's own calculations adjust and calibrate the measurements used by Alcuin. So, in the triangular city problem, 16 houses can be packed into it according to Zolotykh's analysis.

In 1611, Johannes Kepler conjectured that the optimal way to pack spheres of the same size as densely as possible was to pile them up in a pyramidal form. While this can be confirmed easily by trial and error, proving Kepler's conjecture turned out to be as difficult as any of the great problems of mathematics. The proof came in 1994 from American mathematician Thomas Hales using a computer (see Hales 2005).

Packing puzzles are staples of recreational mathematics. Here are two illustrative ones.

Pack 10 equal smaller circles into a larger circle (Figure 10.10).

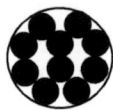

Figure 10.10 Packing Circles in a Circle Puzzle

Pack 15 equal circles into a square (Figure 10.11).

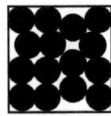

Figure 10.11 Packing Circles in a Square Puzzle

Epilogue

Assembly and partitioning problems are spatial thought experiments, having borne remarkable results in mathematics (Conway and Sloane 1988). Tessellations, which are types of assemblies, are found in Nature. An example is the honeycomb, which is assembled with hexagonal cells. The Greek mathematician Pappus of Alexandria (c. 300–350) maintained that the hexagonal shape was a product of Nature's efficiency—it is the tessellation pattern with the most angles, and thus with the capacity for holding more honey than other geometric figures. Hexagonal wax walls are also quite strong, able to support the bees' load of honey. Pappus' assessment became known as the *honeycomb conjecture*. As German astronomer Johannes Kepler (1611) later remarked (cited in Banks 1999: 19):

What purpose had God in putting these canons of architecture into the bees? Three possibilities can be imagined. The hexagon is the roomiest of the three plane-filling figures (triangle, square, hexagon); the hexagon best suits the tender bodies of the bees; also labour is saved in making walls which are shared by two; labour would be wasted in making circular cells with gaps.

Hexagonal structure also manifests itself in the molecular configuration of snowflakes and ice crystals. It is little wonder, as Banks (1993: 19) puts it, "that currently mathematicians and scientists are devoting much more attention to research on topics advocated by Kepler." Perhaps all this shows is that the brain is an organ that is tuned into the structure of the world (since it rises from it) and comes to understand it better through the science of structure—mathematics.

Explorations

The explorations in this final section include puzzles of various kinds related to the ones devised by Alcuin or else inspired by them subsequently in recreational mathematics.

1. *A Floor Tiling*

A rectangular floor is to be covered with 30 tiles, each measuring 5 × 4. One side of the floor is to be 100 units in length. What will its width be?

2. *Tile Dimensions*

A quadrangular floor measuring 1029 square units in area is covered exactly by 21 tiles. Each of the tiles is a little square. What are the dimensions of each tile?

3. *A Rectangular Floor*

A rectangular floor is designed to have its width be twice its length. It is to be covered exactly by 18 square tiles, measuring 2 × 2 each. What are the dimensions (length, width) of the floor?

4. *A Triangular Roof Top*

A roof top in the form of an isosceles triangle is to be constructed on a building. The façade of the triangle, whose equal sides are 5 units each, and whose base is 6 units, is to be covered by two triangular tiles that fit together perfectly into it, meaning that they are equal. What are the dimensions of those two tiles?

5. *Inserting a Square Carpet in a Circular Room*

Is it possible to insert a square carpet inside a circular room, with its four vertices touching the walls of the circular room?

6. *Carpet Partitioning*

A carpet has 16 little square designs on it. What is the minimal number of cuts needed to produce L-shaped pieces of the carpet consisting of 1, 3, 5, and 7 little squares, respectively (Figure 10.12)?

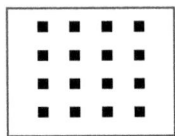

Figure 10.12 Carpet Design

7. *A Linear Path*

A builder wants to design a linear path from one corner of a rectangular floor to the opposite corner. The width of the floor is 20 feet and length 21 feet. What will the length of the path be?

8. *A Circular Design*

A design in the form of a circle is to be drawn on a square floor, with its circumference touching the sides as shown in Figure 10.13.

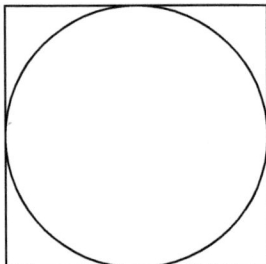

Figure 10.13 Circular Design on Square Floor

If the area of the floor is 25 square feet, what will the area of the circle be?

9. *A Triangular Partitioning*

How could a square floor be partitioned into eight equal triangular sections?

10. *A Puzzle by Henry Dudeney*

In his 1917 puzzle book, *Amusements in Mathematics*, Henry Dudeney presents us with a seemingly impossible puzzle to solve, namely to turn two equal squares into one larger square using four pieces. Can this be done?

Cited Works and Further Reading

Banks, Robert B. (1999). *Slicing Pizzas, Racing Turtles, and Further Adventures in Applied Mathematics*. Princeton: Princeton University Press.

Berlekamp, Elwyn R., Conway, J. Horton, and Guy, Richard K. (1982). *Winning Ways for Your Mathematical Plays*. Boca Raton: CRC Press.

Burkholder, Peter (1993). Alcuin of York's *Propositiones ad acuendos juvenes*: Introduction, Commentary & Translation. *History of Science & Technology Bulletin*, Vol. 1, number 2.

Conway, John Horton (2000). *On Numbers and Games*. Natick, Mass.: A. K. Peters.

Conway, John Horton and Sloane, Neil J. (1988). *Sphere Packings, Lattices and Groups*. Chaim: Springer-Verlag.

Costello, Matthew J. (1988). *The Greatest Puzzles of All Time*. New York: Dover.

Darling, David (2004). *The Universal Book of Mathematics: From Abracadabra to Zeno's Paradoxes*. New York: John Wiley and Sons.

Dudeney, Henry E. (1917). *Amusements in Mathematics*. London: Thomas Nelson and Sons.

Gardner, Martin (1958). *Mathematical Puzzles & Diversions*. Chicago: University of Chicago Press.

Golomb, Sidney W. (1965). *Polyominoes*. New York: Scribners.

Euclid (300 BCE). *The Thirteen Books of Euclid's Elements*, 3 Volumes. New York: Dover, 1956.

Hales, Thomas C. (2001). The Honeycomb Conjecture. *Discrete and Computational Geometry* 25: 1–22.

Hales, Thomas C. (2005). A Proof of the Kepler Conjecture. *Annals of Mathematics, Second Series* 162: 1065–1185.

Kepler, Johannes (1611). *Strena seu de nive sexangular*. Frankfurt: Godefrid Tampach.

Lemon, Don (1890). *Everybody's Book of Illustrated Puzzles*. Miami: Hardpress.

Loyd, Sam (1952). *The Eighth Book of Tan*. New York: Dover.

Netz, Reviel and Noel, William (2007). *The Archimedes Codex: Revealing the Secrets of the World's Greatest Palimpsest*. London: Weidenfeld & Nicholson.

O'Connor, John J. and Robertson, Edmund F. (2012). *Propositiones ad acuendos juvenes*. https://mathshistory.st-andrews.ac.uk/HistTopics/Alcuin_book/.

Penrose, Roger (1974). The Role of Aesthetics in Pure and Applied Mathematical Research. *Bulletin of the Institute of Mathematics and its Applications* 10: 266–271.

Slocum, Jerry, Botermans, Jacob, Gebhardt, Dieter, Ma, Monica, Ma, Xiaohe, Raizer, Harold, Sonneveld, Dic, and van Splumteren, Carla (2004). *The Tangram Book: The Story of the Chinese Puzzle with Over 2000 Puzzles to Solve*. New York: Sterling.

Takagi, Shigeo (1999). Japanese Tangram: The Sei Shonagon Pieces. In: Elwin Berlekamp and Tom Rodgers (eds.), *The Mathemagician and Pied Puzzler: A Collection in Tribute to Martin Gardner*, 97–98. Natick, Mass.: A. K. Peters.

Vorderman, Carol (1996). *How Math Works*. Pleasantville: Reader's Digest Association.

Zolotykh, Nikolai Yu (2013). Alcuin's *Propositiones de Civitatibus*: The Earliest Packing Problems. *arXiv*: 1308.0892.

11
Concluding Remarks

Relevance of the *Propositiones*

It is a cliché to say that, since the dawn of civilization, mathematics has played a vital role in human life throughout the world. But this does not diminish the verity of that phrase. Mathematics provides powerful ways to think about, represent, and act upon, the world, even though its symbols, formulas, methods, and various technical conventions may seem, on the surface, to be abstruse and disconnected from the issues of everyday life. Since antiquity, knowledge of mathematics has always been considered to be a requisite of the "cultured" person, to be learned by everyone. Alcuin's *Propositiones* provides a snapshot of what such knowledge is about, through 50 plus ingenious problems, which put on display how basic mathematics works, and what it allows us to do, practically and cognitively. Although designed pedagogically, Alcuin's text is not just a school textbook in the ordinary sense of the term, but rather a collection of ingenious puzzles that show what mathematical method is about in a nutshell.

Alcuin's book has also been the source of several important ideas that stand out in the history of mathematics, some of which have had an impact on human life. In a relevant 1998 article, David Singmaster aptly remarked that the *Propositiones* is a "major landmark in the history of mathematics in general." As discussed in several *Annotation* sections of this book, there are novel ideas mixed in with established ones in Alcuin's book—ideas that would blossom theoretically after Alcuin. Singmaster put it as follows: "Some of the problems have had lengthy developments since the time of Alcuin, occurring in almost every arithmetic-algebra text down to the modern day."

The Importance of Recreational Mathematics

What goes on in our minds as we solve math problems, such as those by Alcuin? As Robert Sternberg (1985) showed with several experimental studies a while back, solving puzzles involves the simultaneous utilization of three thinking modes. The first one, which he called selective encoding, refers to the process of selecting information that is relevant to the task at hand, while discarding information that is not. In so doing we "squeeze out" hidden ideas from the puzzle. The second, called selective comparison, entails making hunches, in order to draw non-obvious comparisons between the ideas in the puzzle and what we know already about them. And the third, named selective combination, involves connecting the ideas in order to envision a singular solution that allows us to grasp the meaning of the hidden ideas. This "triarchic model" seems to be an appropriate one for characterizing how many of Alcuin's problems are solved. First, we must extract the relevant information from them, leaving aside extraneous information and ignoring impossible or erroneous computations. Second, for many of the problems, we need to make hunches about the pattern they conceal. Third, we must put the hunches together cohesively, so as to be able to solve the problem.

In a comprehensive treatment of the origins and evolution of puzzles, mathematical and otherwise, Helene Hovanec (1978: 10) observes that puzzles are felt to be inherently challenging because they "simultaneously conceal the answers yet cry out to be solved." It is this "crying out to be solved" that characterizes the overall appeal of Alcuin's recreational mathematics.

What is Math?

Consider the following puzzle by Henry E. Dudeney, published in the *Strand Magazine* (volume 77, 1929):

> Arrange all the ten digits in three arithmetical sums, employing three of the four operations of addition, subtraction, multiplication, and division, and using no signs except the ordinary ones implying those operations.

Dudeney's solution is the following one. Note that all the digits are used, including 0 (in the digit "20"):

$$7 + 1 = 8$$
$$9 - 6 = 3$$
$$4 \times 5 = 20$$

In its own ingenious way, this puzzle bears the main features of mathematical thinking. At first, we are given a challenge that seems to either defy a solution or at least requires quite of bit of cogitation to solve. By imagining and trying out certain possibilities, using background knowledge of mathematics, eventually a solution appears, as if by magic. This type of challenging, yet ultimately self-rewarding, creative process is the essence of mathematics as an art of thinking. It is this art that the *Propositiones* exemplifies. Mathematics is the science of hidden patterns, be it numerical, geometrical, or graphical. It is the play on this pattern that establishes Alcuin's work as a major one in recreational mathematics.

Some of Alcuin's problems show up in different guises in other parts of the world at other times, suggesting that their ideation transcends cultural spaces and specific eras. For example, similar ones appear in the third-century *Bhakshali Manuscript*, discovered in northwest India in 1881. This has suggested to some that Alcuin may have borrowed some of his problems from other cultures. But this possibility is highly unlikely, since he would have had to know other languages in an era where learning foreign languages was not common; moreover, as discussed in this book, a number of his problems are original, with no known precedents. Whoever composed the *Propositiones*, in the end, the book shows that mathematics is, as the mathematician G. H. Hardy described it in 1940, a "creative art."

Cited Works and Further Reading

Singmaster, David (1998). The History of Some of Alcuin's *Propositiones*. In: P. L. Butzer, H. Th. Jongen, and W. Oberschelp (eds.), *Charlemagne and His Heritage 1200 Years of Civilization and Science in Europe*, Vol. 2, pp. 11–29. Brepols: Turnhout.

Hardy. G. H. (1940). *A Mathematician's Apology.* Cambridge: Cambridge University Press.

Hovanec, Helene (1978). *The Puzzlers' Paradise: From the Garden of Eden to the Computer Age.* New York: Paddington Press.

Sternberg, Robert J. (1985). *Beyond IQ: A Triarchic Theory of Human Intelligence.* New York: Cambridge University Press.

Answers and Solutions
to the Explorations

Chapter 1

[Note that the "A" in the numeration of figures stands for "Answers."]

1. *Answer*: cow = 52 ½ gold coins; the goat = 2 ½ gold coins

Solution

The solution can be broken down as follows:

- Since the total cost of the cow and the goat is 55 coins, and the cow cost 50 coins more than the goat, this means that the cow itself cost more than 50 coins.
- Saying that the cow cost 50 coins more than the goat is equivalent to saying that the goat cost 50 coins less than the cow.
- Let us assume that the cow cost 51 coins. Then, the goat would have cost 1 gold coin, which is 50 coins less.
- But 51 + 1 = 52, not 55, the whole cost.
- So, let us try 52 coins for the cow. The goat would then have cost 2 coins, which is 50 coins less.
- But again, 52 + 2 = 54, not 55.
- Let us try 53 for the cow. In this case, the goat would have cost 3 coins, which is 50 coins less.
- This time the total cost would be 53 + 3 = 56, which is one more than required.
- So, logically, the cow cost a little more than 52 and less than 53. It thus cost 52 ½ coins. This means that the goat would have cost 2 ½ coins, which is 50 coins less.
- Together, they add up to 52 ½ + 2 ½ = 55, which is the correct amount.

Algebra could be used instead to solve the problem:

- Let x be the cost of the goat. This means that $(x + 50)$ is the cost of the cow, which stands for "50 gold coins more than the unknown cost, x, of the cow."
- The two add up to 55 gold coins. The relevant equation is, therefore:

$$x + (x + 50) = 55$$
$$2x + 50 = 55$$
$$2x = 5$$
$$x = 2\ ^1/_2, \text{ the cost of the goat}$$
$$x + 50 = 2\ ^1/_2 + 50 = 52\ ^1/_2, \text{ the cost of the cow.}$$

Using either arithmetical or algebraic thinking to solve the problem, the goat cost 2 ½ coins and the cow, costing 50 coins more, cost 52 ½ coins. Together, they add up to 55:

$$2 \tfrac{1}{2} + 52 \tfrac{1}{2} = 55$$

2. *Answer*: see solution

Solution

In order for Achilles to surpass the tortoise, he must first reach the halfway point, which is the tortoise's starting point. But when he does, the tortoise will have moved forward a little bit. Achilles must then reach the tortoise's new point before surpassing it. When he does, however, the tortoise has again moved a little bit forward, which Achilles must also reach again, and so on ad infinitum. In other words, although the distance between Achilles and the tortoise will get smaller and smaller (in fact, infinitesimally so), Achilles will never surpass the tortoise.

We can model the situation with a line graph (A = Achilles, T = Tortoise). At the start, the linear path to the finish point can be represented as shown (Figure A1.1). Note that subscripts are used to indicate relative position on the graph. So, A_0 = initial position for Achilles, A_1 = first position of Achilles after moving away from the start, A_2 = second position of Achilles from the start, and so on. The same notation is used for the tortoise.

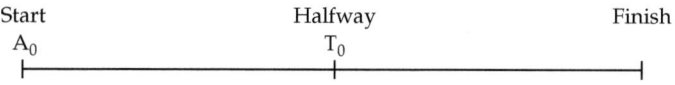

Figure A1.1 Initial Position of Achilles and the Tortoise

After Achilles reaches the halfway point, the tortoise would have gone a bit forward (Figure A1.2).

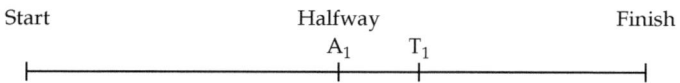

Figure A1.2 First Positions after the Initial State

After Achilles reaches T_1, represented by A_2, the tortoise will have gone forward a bit to position T_2 (Figure A1.3).

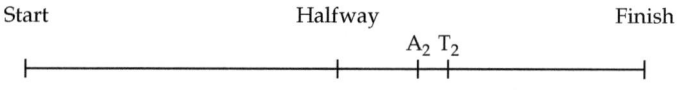

Figure A1.3 Second Positions after the Initial State

The distance between A_n and T_n will get smaller and smaller, but the two positions will never coincide. Of course, in reality, Achilles, being the faster runner, will surpass the tortoise, but explaining why he does in terms of the paradox soon after became a major debate in philosophy, science, and mathematics, leading over time to the theory of limits and the calculus. Zeno's paradoxes portray movement in terms of discrete points. To resolve them, a distinction between discrete and continuous is required—a distinction made with the calculus much later.

3. *Answer*: 15 miles

Solution

The bike riders were 20 miles before they met. Since both were traveling at 10 miles per hour toward each other, they would therefore meet halfway: that is, after they had each covered 10 of the 20 miles. At 10 miles per hour, the riders covered this distance in 1 hour. Since the fly went back and forth at 15 miles per hour during that 1 hour, it covered a total distance of 15 miles—that is, the fly took the same time as the bike riders to cover the 15 miles, namely one hour. The fly did it in a zig-zag fashion, of course, not a straight line.

4. *Answer*: 25 rungs

Solution

This is a number line puzzle in disguise. In the beginning, we do not know what rung the firefighter is on, except that it is the middle rung. So, we can label her starting rung as 0: that is, we can consider the middle rung to be the zero point on the number line. Since it is the middle rung, there will be as many rungs above it as there are below it, so as to make an actual ladder. The firefighter starts by going up three rungs from the 0 rung, which means that she reached the third rung above (Figure A1.4).

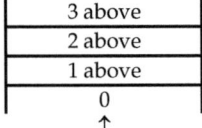

Figure A1.4 Up 3 from 0

She then stepped down five rungs. This means that from rung 3 above, she went down five steps to rung number 2 below the middle rung (Figure A1.5).

$$\downarrow$$

3 above
2 above
1 above
0
1 below
2 below

Figure A1.5 Down 5 to 2 Below

Next, the firefighter climbed up seven rungs from rung 2 below, ending up at rung 5 above (Figure A1.6).

| 5 above |
| 4 above |
| 3 above |
| 2 above |
| 1 above |
| 0 |
| 1 below |
| 2 below |
| ↑ |

Figure A1.6 Up 7 to 5 Above

Finally, the firefighter climbed up another seven rungs (from rung 5) to the roof. This means that she climbed to rung 12 above her starting point (Figure A1.7).

| 12 above |
| 11 above |
| 10 above |
| 9 above |
| 8 above |
| 7 above |
| 6 above |
| 5 above |
| 4 above |
| 3 above |
| 2 above |
| 1 above |
| 0 |
| 1 below |
| 2 below |
| ↑ |

Figure A1.7 Up 5 to 12 Above

Rung 12 above is the top part of the ladder. So, let us complete the ladder. We know that it has 12 rungs above the middle rung, so a complete ladder will also have 12 rungs below it (Figure A1.8).

The ladder thus has 12 rungs above the 0 rung, 12 below it, and the 0 rung itself. This makes 25 rungs in total. To use the number line as a model, it can be seen that the number of rungs above the middle rung is equivalent to the positive integers to the right of 0 on the line, and the number of rungs below it is equivalent to the negative numbers to the left of 0 (Figure A1.9).

5. *Answer*: 10 trains

Solution

Let us suppose that we get on a train at the New York station at 12:00 noon. Five hours later, at 5:00 p.m., our train arrives at the Washington station. Now, let us draw a diagram charting the relative positions of the trains that were on their way from Washington to New York during those five hours, keeping in mind that the Washington-to-New York trains leave on the half hour.

12 above
11 above
10 above
9 above
8 above
7 above
6 above
5 above
4 above
3 above
2 above
1 above
0 = middle rung
1 below
2 below
3 below
4 below
5 below
6 below
7 below
8 below
9 below
10 below
11 below
12 below

Figure A1.8 Complete Ladder

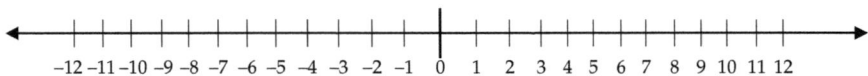

Figure A1.9 Number Line Model of the Warehouse Fire Problem

At 5:00 p.m. at the Washington station, there is a train ready to leave for New York. Let us call it A (Figure A1.10).

Figure A1.10 Location of Train A

Now, the train that left a half hour earlier from Washington will find itself at a distance B at 4:30 p.m. (Figure A1.11).

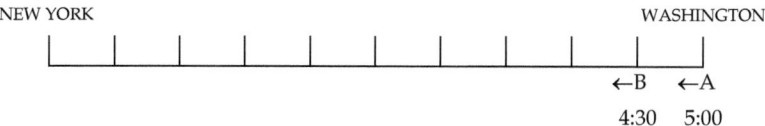

Figure A1.11 Location of Train B

Continuing to reason in this way, we can then complete the diagram showing the relative locations of all the trains from Washington bound for New York which left at half-hour intervals between 12:00 p.m. and 5:00 p.m. (Figure A1.12).

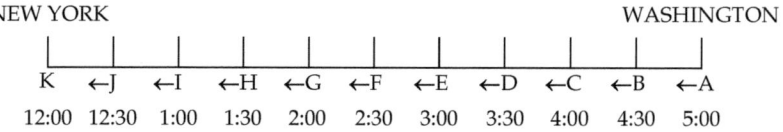

Figure A1.12 Position of Trains A to K

Now, when we left the New York station at 12:00 noon, we obviously missed passing the 12:00 o'clock K-train from Washington, because it was already in the station. From the diagram above, however, it can be seen that we passed all the others, making 10 trains in all:

(1) 12:30 → J-train
(2) 1:00 → I-train
(3) 1:30 → H-train
(4) 2:00 → G-train
(5) 2:30 → F-train
(6) 3:00 → E-train
(7) 3:30 → D-train
(8) 4:00 → C-train
(9) 4:30 → B-train
(10) 5:00 → A-train (in the station as it is leaving)

6. *Answer*: 222

Solution

Here is how it can be broken down:

- The best approach to solving this problem is to work backward.
- The only three identical digits which, when added together, add up to "3" are "1 + 1 + 1."
- So, "111" is half of the number of coins, meaning that the number was "222."
- In fact, if we divide 222 by half we get 111, whose digits add up to 3.

7. *Answer*: 3 mph (miles per hour)

Solution

One might assume that the average speed is calculated simply by adding the hiker's two rates—2 mph and 6 mph—together (= 8 mph) and then dividing by 2. This leads to the incorrect answer of 4 mph.

First, let us recall the formula which shows that the distance covered is equal to the rate multiplied by the time taken:

$$\text{Distance} = \text{Rate} \times \text{Time}$$

$$D = R \times T$$

Now, we can calculate the time Danielle took to hike uphill. For the sake of convenience, let us assume that the distance uphill (and downhill, of course) is 1 mile. We could use any other distance, but the reasoning and result would be the same. We are told that her rate uphill is 2 mph. So, her *time uphill* is calculated as shown:

$$D = 1, \ R = 2$$

$$D = R \times T$$

$$1 = 2 \times T$$

$$T = 1/2$$

Thus, it would take Danielle 1/2 hour to climb a 1-mile hill. Now, given that her rate downhill is 6 mph, her *time downhill* is calculated as follows:

$$D = 1, \ R = 6$$

$$D = R \times T$$

$$1 = 6 \times T$$

$$T = 1/6$$

So, it would take Danielle 1/6 hour to descend a 1-mile hill. Now, her total time for the entire trip is *time uphill* + *time downhill*:

$$1/2 + 1/6 = 4/6 = 2/3 \text{ hour}$$

The total distance Danielle covered is 2 miles—1 mile up + 1 mile down. Therefore, her overall rate can be calculated as follows:

$$D = 2, \ T = 2/3$$

$$D = R \times T$$

$$2 = R \times 2/3$$

$$R = 2 \div 2/3 = 2 \times 3/2 = 3 \text{ mph}$$

Danielle's average rate of speed, which has been calculated by taking into account the overall distance she covered (= 2 miles) and the overall time she took (= 2/3 hour), is 3 mph, not 4 mph.

8. *Answer*: 7 dimes and 13 nickels

Solution

Let us start by assuming that Jim has an equal number of dimes and nickels in his pocket—10 of each. One dime is equal to $0.10 of a dollar, and one nickel is worth

$0.05 of a dollar. So, 10 dimes are worth $10 \times 0.10 = \$1.00$, and 10 nickels are worth $10 \times 0.05 = \$0.50$.

Scenario 1
10 dimes = $1.00
10 nickels = $0.50
Total   = $1.50

The total is too high, because Jim has only $1.35 in his pocket. So, let us try a different combination. If a dime is taken away from the 10 dimes in scenario 1, then the number of nickels must be increased by one—because Jim has 20 coins in his pocket. Below is what 9 dimes and 11 nickels add up to in money value:

Scenario 2
9 dimes   = $0.90
11 nickels = $0.55
Total   = $1.45

This total is still greater than $1.35. So, let us reduce the number of dimes in Jim's pocket by one more (8 dimes), while increasing the nickels to one more (12 nickels):

Scenario 3
8 dimes   = $0.80
12 nickels = $0.60
Total   = $1.40

This total is still too high, but we are getting closer to the required total of $1.35. So, let us see what happens when the number of dimes in Jim's pocket is decreased again by one (7 dimes), while the nickels are simultaneously increased by one (13 dimes):

Scenario 4
7 dimes   = $0.70
13 nickels = $0.65
Total   = $1.35

As can be seen, scenario 4 contains the solution—Jim has 7 dimes and 13 nickels in his pocket.

9. *Answer*: Amber 9 days, Beatrice 18 days

Solution

It takes Amber and Beatrice six days to complete the job together. This means that in one day they will complete 1/6 of the job. To figure out how many days each takes to do the job working alone, we can use the same reasoning of daily work rate for each person. Let us rephrase the problem in strict arithmetical terms as follows:

What two fractions (representing the daily work rates of each person) add up to 1/6, with one fraction twice the value of the other fraction?

Arithmetically, the smaller fraction will have a denominator that is twice the denominator of the larger fraction. To see this, consider a few examples:

- 1/2 is twice 1/4, which can be shown by multiplying 1/4 by 2:

$$1/4 \times 2 = 1/2$$

- 1/3 is twice 1/6, which can be shown by multiplying 1/6 by 2:

$$1/6 \times 2 = 1/3$$

- 1/4 is twice 1/8, which can be shown by multiplying 1/8 by 2:

$$1/8 \times 2 = 1/4$$

We must come up with two fractions that add up to 1/6, with one having a denominator that is twice the other denominator and each one less in value than 1/6. A little trial and error will show that only 1/9 and 1/18 will work:

$$1/9 + 1/18 = 1/6$$

Thus, 1/9 means that Amber takes nine days working alone and 1/18 means that Beatrice takes 18 days working alone. The problem can also be solved using algebra. Let the number of days Amber needs to finish the job alone be x. Her daily work rate is thus $1/x$. Beatrice takes twice as long, so she will finish the job in $2x$ days, meaning that her work rate is $1/2x$. Working together they complete 1/6 of the job in a day:

$$1/x + 1/2x = 1/6$$
$$6/x + 6/2x = 1$$
$$6/x + 3/x = 1$$
$$6 + 3 = x$$
$$x = 9$$
$$2x = 2 \times 9 = 18$$

This tells us that Amber will require nine days to complete the job working alone and Beatrice 18 days.

10. *Answer*: 6

Solution

Arithmetically, the puzzle can be framed simply as follows: Which combination of scores on the target adds up to one hundred? This makes the solution a little easier to envision. The young lady used six arrows because only the following six scores, when added together, add up to 100:

$$17 + 17 + 17 + 17 + 16 + 16 = 100$$

No other combination of the numbers that represent the scores—16, 17, 23, 24, 39, and 40—will add up to 100.

Chapter 2

1. *Answer*: 59 pigs

Solution

Counting pigs by ones, twos, threes, and so on is equivalent to dividing the number of pigs into smaller sets of one pig, two pigs, three pigs, and so on. Now, we must identify the number of pigs between 50 and 60 which, when divided by 3, gives a remainder of 2, which is equivalent to saying that there are two pigs left over, and when divided by 5, gives a remainder of 4, which is equivalent to saying that there are four pigs left over.

First, we divide the numbers between 50 and 60 by 3, identifying those that leave a remainder of 2:

$$50 \div 3 = 16, \text{ remainder} = 2$$
$$51 \div 3 = 17, \text{ no remainder}$$
$$52 \div 3 = 17, \text{ remainder} = 1$$
$$53 \div 3 = 17, \text{ remainder} = 2$$
$$54 \div 3 = 18, \text{ no remainder}$$
$$55 \div 3 = 18, \text{ remainder} = 1$$
$$56 \div 3 = 18, \text{ remainder} = 2$$
$$57 \div 3 = 19, \text{ no remainder}$$
$$58 \div 3 = 19, \text{ remainder} = 1$$
$$59 \div 3 = 19, \text{ remainder} = 2$$
$$60 \div 3 = 20, \text{ no remainder}$$

We can now see that the only numbers between 50 and 60, which, when divided by 3, will leave a remainder of 2 are the numbers 50, 53, 56, and 59. We can now discard the others, proceeding to determine which number, 50, 53, 56, or 59, will leave a remainder of 4 when it is divided by 5:

$$50 \div 5 = 10, \text{ no remainder}$$
$$53 \div 5 = 10, \text{ remainder} = 3$$
$$56 \div = 11, \text{ remainder} = 1$$
$$59 \div 5 = 11, \text{ remainder} = 4$$

As can be seen, 59 pigs is the answer. And, in fact, when we count 59 pigs 3 at a time, we will get 2 left over, and when we count them 5 at a time, we will get 4 left over.

2. *Answer*: see solution

Solution

We start by dividing the six balls into two sets of three balls each.

Weighing 1

- For the first weighing, we put three balls on each pan—three on the left pan and three on the right pan.
- The pan that goes up contains the ball that weighs less, but we do not yet know which one of the three.
- So, we retain the set containing the suspect ball, discarding the ones on the other pan.
- We select any two of the three balls to weigh, putting the third ball aside.

Weighing 2

- We put each one of the two randomly selected balls on a separate pan—one on the left pan and one on the right pan.
- If the pans balance, then the suspect ball is the one we set aside. If they do not, then the pan that goes up will contain it.
- Either way, the second weighing will identify the ball that weighs less.

Increasing the number of balls does not change the reasoning involved—it becomes more complex, not more difficult or different. For example, consider an odd number of balls to be weighed, such as 7. We are asked again to identify the one ball that weighs less than the others. We start by dividing six balls into two sets of three balls each, and putting the seventh one aside.

Weighing 1

- We put three balls on each pan—three on the left pan and three on the right pan.
- If they balance, then we have found the culprit ball by elimination—the one we had put aside.
- If that happens by luck, we would have needed just one weighing. But mathematicians do not rely on luck, but on hypothesis thinking, which in this case involves a worst-case scenario—that is, we should assume that it did not happen. The solution is then exactly the same as it was for the previous problem.
- The pan that goes up contains the ball that weighs less.
- We retain the set containing the suspect ball, discarding the ones on the other pan.
- We select any two of the three balls, putting the third ball aside.

Weighing 2

- We put each one of the two randomly selected balls on a separate pan—one on the left pan and one on the right pan.
- If the pans balance, then the suspect ball is the one we set aside. If they do not, then the pan that goes up will contain it.
- Either way, the second weighing will identify the ball that weighs less.

3. *Answer*: four

Solution

One might conclude at first that six weights of "1," "2," "4," "8," "16," and "32" pounds would be needed. The reasoning might go somewhat as follows.

- We could weigh "1 pound" of sugar by putting the "1-pound weight" on the left pan, pouring sugar into the right pan until the pans balance.
- We could weigh "2 pounds" of sugar by putting the "2-pound" weight on the left pan, pouring sugar on the right pan until the pans balance.
- We could weigh "3 pounds" of sugar by putting the "1-pound" and the "2-pound" weights on the left pan, pouring sugar on the right pan until the pans balance.
- And so on.
- In this way, we could weigh any number of integral (whole-number) pounds of sugar from "1 pound" to "40 pounds."

But this is certainly not the *least number* of weights, which the puzzle asks us to determine. Since we can put the weights on both pans of the scale, the weighing can be done, Bachet indicates, with only four weights of "1," "3," "9," and "27" pounds. The reason for this is that placing a weight on the right pan, along with the sugar, is equivalent to taking its weight away from the total weight on the left pan. For example, if "2 pounds" of sugar are to be weighed, we would put the "3-pound" weight on the left pan and the "1-pound" weight on the right pan. We will get a balance when we pour the missing "2 pounds" of sugar on the right pan. This type of reasoning applies to all the pounds of sugar to be weighed.

The four weights are, upon closer scrutiny, powers of "3":

$$1 = 3^0$$
$$3 = 3^1$$
$$9 = 3^2$$
$$27 = 3^3$$

The choice of these weights works because, as it turns out, each of the whole numbers from "1" to "40" (= the required weights) are integers that can be expressed with powers of "3":

$$1 \text{ pound} = 3^0 = 1$$
$$2 \text{ pounds} = (3^1 - 3^0) = (3 - 1) = 2$$
$$3 \text{ pounds} = 3^1 = 3$$
$$4 \text{ pounds} = (3^1 + 3^0) = (3 + 1) = 4$$
$$5 \text{ pounds} = 3^2 - (3^1 + 3^0) = 9 - (3 + 1) = 9 - 4 = 5$$
$$\dots$$
$$40 \text{ pounds} = 3^3 + 3^2 + 3^1 + 3^0 = 27 + 9 + 3 + 1 = 40$$

The chart below summarizes how the weighing process is carried out:

Sugar Amount	Weights Put on the Left Pan	Weights Put on the Right Pan
1 pound	A one-pound weight	No weight needed, just sugar, which will weigh one pound, which is the amount required to make the scales balance.

Continued

Continued

Sugar Amount	Weights Put on the Left Pan	Weights Put on the Right Pan
2 pounds	A three-pound weight	A one-pound weight plus sugar, which will weigh 2 pounds to make the scales balance.
3 pounds	A three-pound weight	No weight needed, just sugar, which will weigh 3 pounds to make the scales balance.
4 pounds	A one-pound weight and a three-pound weight	No weight needed, just sugar, which will weigh 4 pounds to make the scales balance.
5 pounds	A nine-pound weight	A one-pound and a three-pound weight plus sugar, which will weigh 5 pounds to make the scales balance.
. . .	. . .	. . .
40 pounds	A one-pound weight, a three-pound weight, a nine-pound weight, and a 27-pound weight	No weight needed, just sugar, which will weigh 40 pounds.

4. *Answer*: 2/3

Solution

Let "B" and "W-1" stand respectively for the black or white counter that *might* be in the bag at the start, and "W-2" for the white counter added to the bag. Now, removing a white counter from the bag entails three equally likely hypothetical combinations of two counters, one inside and one outside the bag:

Inside the Bag	Counter Added	Bag Content	Counter Drawn
(1) W - 1	W - 2	W - 1, W - 2	W - 2
(2) W - 1	W - 2	W - 1, W - 2	W - 1
(3) B	W - 2	B, W - 2	W - 2

The chance of drawing a white counter on the second draw is two out of three. This can be explicated as follows:

- In hypothetical combination (1), there was a white counter in the bag, W-1, to which the external white counter, W-2, was added. The white counter drawn out will be one of the two white ones inside. Let us assume that it was the one put into the bag, namely, W-2.
- Hypothetical combination (2) is the converse of (1). Again, there was a white counter in the bag, W-1, to which W-2 was added, making two white counters in the bag. This time, the white counter drawn out is the one that was originally in the bag, W-1.
- In hypothetical combination (3), there was a black counter, B, in the bag to which the external white counter, W-2, was added, making a black and white counter in the bag. The white counter drawn out is the one that was put in

the bag, W-2, since there was no white counter originally in it. The counter that remains in the bag is a black one.
- In two of the three hypothetical scenarios, Carroll observed, a white counter remains in the bag. So, the chance of drawing a white counter on the second draw is two out of three.

5. *Answer*: It cannot be done.

Solution

The problem asks us, in effect, to find five consecutive odd numbers that add up to 228. But this is impossible, since the addition of five consecutive odd numbers will equal an odd number, not an even one.

The addition of consecutive odd numbers can be represented as follows:

$$(2n + 1) + (2n + 3) + (2n + 5) + (2n + 7) + (2n + 9) = 10n + 25$$

No matter what value we assign to n, the result will always be an odd number. For example, if $n = 3$, then $10n + 25 = 30 + 25 = 55$. So, $10n + 25$ does not represent an even number, such as 228, the number of pigs.

6. *Answer*: see solution

Solution

To accommodate the new guest the proprietor does the following:

- He moves the person previously occupying the first room, "N_1," into the next room, "N_2."
- He moves the person who had been occupying room "N_2" into the next room "N_3."
- He moves the person who had been occupying room "N_3" into the next room "N_4."
- And so on, ad infinitum (since the hotel is an infinite hotel).
- The new guest is then put into room "N_1," which became free as the result of these transpositions.

7. *Answer*: see solution

Solution

In the case of an infinite number of guests wanting rooms in the same infinite hotel, the proprietor does the following:

- He moves the person occupying "N_1" to the room occupied by "N_2".
- As a result "N_1" becomes free.
- He then moves the guest occupying "N_2" to "N_4", and, in general, a guest occupying room n will move to an even-numbered room $2n$ times.
- As a result of such transpositions, all odd-numbered rooms become free and the infinite number of new guests can easily be accommodated in them, while the others now occupy the even-numbered rooms.
- The reason why an infinite number of guests can be accommodated is because there is an infinity of even numbered rooms and an infinity of odd-numbered rooms which can be put into a one-to-one-correspondence.

8. *Answer*: 8^2

Solution

The solution can be broken down as follows:

- Number of chickens: the original three each produced three of their own, or $3 \times 3 = 3^2$. So, the number of chickens in the farm was the original $3 + 3^2 = 3 + 9 = 12$.
- Number of pigs: the original four each produced four of their own, or $4 \times 4 = 4^2$. So, the number of pigs in the farm was the original $4 + 4^2 = 4 + 16 = 20$.
- Number of cows: the original five each produced five of their own, or $5 \times 5 = 5^2$. So, the number of cows in the farm was the original $5 + 5^2 = 5 + 25 = 30$.
- Total number of animals: $12 + 20 + 30 = 62$.
- Adding 2 to this number: $62 + 2 = 64 = 8^2$

9. *Answer*: 9, 2, 2

Solution

There are various possibilities of three numbers, standing for the children's ages, which, when multiplied together, give 36. But only 9, 2, 2, and 6, 6, 1 add up to the sum of 13, which is thus the number on the gate:

$$Product\ of\ their\ ages:\ 9 \times 2 \times 2 = 36 \text{ or } 6 \times 6 \times 1 = 36$$
$$Sum\ of\ their\ ages:\ 9 + 2 + 2 = 13 \text{ or } 6 + 6 + 1 = 13$$

No other integer triplet will satisfy these two conditions; that is, no other set of three numbers will produce a product of 36 and a sum of 13. When the resident stated that her "oldest" child had measles, the census taker realized that the triplet that allowed for one, and only one, oldest child was 9, 2, 2. Therefore, these are the ages of the children. In 6, 6, 1, there are two, not one, older children.

10. *Answer*: six

Solution

We must assume the worst-case scenario as discussed in the Annotations:

(1) *First draw*: Assume that the first ball we draw out is the yellow one. This initiates the worst-case scenario because there is no matching yellow ball in the box.
(2) *Second draw*: The second ball we draw out will be one of the other balls. Let us suppose that we draw a white ball (it could be any other colored ball).
(3) *Third draw*: After drawing out the white ball, under the worst-case scenario, the next one will be of a different color, say, black.
(4) *Fourth draw*: The one after that will be of yet another color, say green.
(5) *Fifth draw*: And finally, the one after that will be blue. At this point, we have drawn out five balls, all of different colors, including the yellow one.
(6) *Sixth draw*: Now, the next ball we draw out will necessarily match one of the colors, other than the yellow one, of course. If we draw one of the remaining white balls, it will match the white ball outside; if we draw a remaining black ball, it will match the black ball outside; if we draw a remaining green

ball, it will match the green ball outside; and if we draw a remaining blue ball, it will match the blue ball outside. Thus, the sixth draw will produce a match for sure.

Chapter 3

1. *Answer*: 12 feet, 12 feet, 36 feet

Solution

Let each of the two lateral sides of the pen be x feet. Its front side will be three times this, or $3x$. The dimensions of the three sides can be shown as follows, noting that its fourth side is attached to the larger barn (Figure A3.1).

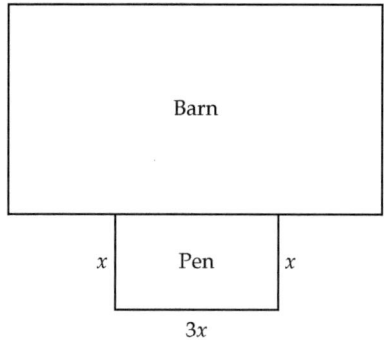

Figure A3.1 Dimensions of the Pen

The total number of feet required to fence in the pen is, therefore:

$$x + x + 3x = 5x \text{ feet}$$

We are told that the farmer has 60 feet of wire to cover the fencing. Therefore:

$$5x = 60$$
$$x = 60/5$$
$$x = 12$$

This is the length of one lateral side (Figure A3.1). So, the two lateral sides are 12 feet each, adding up to $12 + 12 = 24$ feet. Now, $3x$ stands for the length of the front side (Figure A3.1), therefore, $3x = 3 (12) = 36$ feet. These are the dimensions of the pen: 12 feet, 12 feet, 36 feet.

2. *Answer*: 2 feet all around

Solution

Let the dimension of the walk be x feet all around the pool, which is 9 feet-by-4 feet (Figure A3.2).

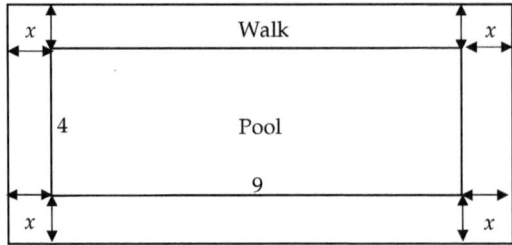

Figure A3.2 Dimensions of the Pool and the Walk

Note that a larger rectangle is formed including the rectangular pool, created by adding the walk. The width of that rectangle is equal to the width of the pool (4 feet), plus $2x$, and its length is equal to the length of the pool (9 feet), plus $2x$ (Figure A3.3).

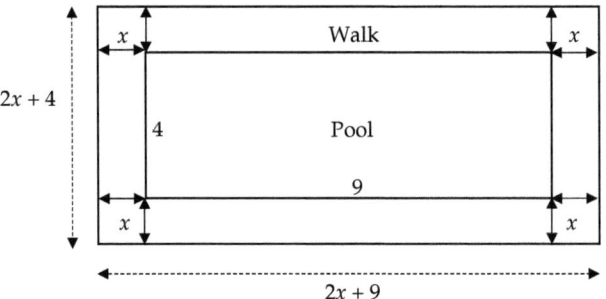

Figure A3.3 Dimensions of the Larger Rectangle

Since we now know the length and width of the larger rectangle, we can compute its area as follows (quadratic equations will be discussed in Chapter 5):

$$\text{Area} = \text{length} \times \text{width}$$
$$(2x + 9) \times (2x + 4) = 4x^2 + 26x + 36$$

The area of the fish pool is $9 \times 4 = 36$. So, the area of the walk is determined as follows:

Area of larger rectangle minus the area of the fish pool:

$$(4x^2 + 26x + 36) - 36 = 4x^2 + 26x$$

We are told that the contractor can only afford to make this 68 feet. So:

$$4x^2 + 26x = 68$$
$$4x^2 + 26x - 68 = 0$$

Using the method of factorization:

$$(2x + 17)(2x - 4) = 0$$

Therefore, either:

$$2x + 17 = 0$$

or:

$$2x - 4 = 0$$

Consider the first possibility:

$$2x + 17 = 0$$
$$2x = -17$$
$$x = -17/2$$

This solution is not possible because it is a minus length.
  Consider the other possibility:

$$2x - 4 = 0$$
$$2x = 4$$
$$x = 2$$

The contractor should make his walk 2 feet all around.

3. *Answer*: 7 inches-by-14 inches

Solution

Draw a rectangular cardboard, letting x stand for its width and $2x$ for its length, which is twice as much (Figure A3.4).

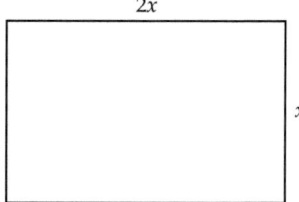

Figure A3.4 Dimensions of the Cardboard

  Next, draw 2-inch squares in each corner, showing, with dash lines, where two of their sides will be cut out so that the sides of the cardboard can be then turned up evenly to make a box. After the cut-outs, note that the width of the cardboard sheet will be $(x - 4)$ inches, and the length $(2x - 4)$ inches (Figure A3.5).

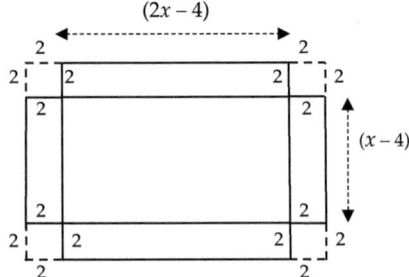

Figure A3.5 Dimensions of the Cardboard after the Cut-Outs

Now, turning up the sides of the cardboard after the cut-outs, we get the required box, with length $(2x-4)$, width $(x-4)$, and height 2 inches (Figure A3.6).

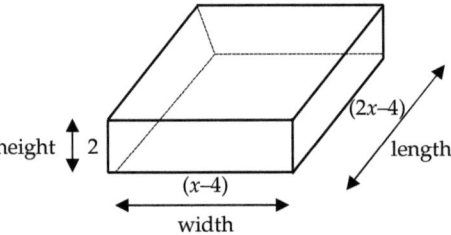

Figure A3.6 Rectangular Box

The volume of the rectangular box is given as 60 cubic inches. Therefore:

$$\text{Volume} = \text{length} \times \text{width} \times \text{height} = 60$$
$$(2x-4)(x-4) \times 2 = 60$$
$$(2x^2 - 12x + 16) = 60/2$$
$$2x^2 - 12x + 16 = 30$$
$$2x^2 - 12x - 14 = 0$$
$$x^2 - 6x - 7 = 0$$

Using the method of factorization:

$$(x-7)(x+1) = 0$$

Therefore, either:

$$(x-7) = 0$$

or:

$$(x+1) = 0$$

Consider the second possibility:

$$x + 1 = 0$$
$$x = -1$$

This solution is not possible because it is a minus length.
   Consider the first possibility:

$$x - 7 = 0$$
$$x = 7$$

Recall that x is the original width of the cardboard, while $2x$ was the original length. So, the width was 7 and the length 14, which is twice the width. The dimensions of the original piece of cardboard were 7 inches-by-14 inches.

4. *Answer*: see solution

Solution

Cut the figure in a zig-zag fashion, producing a staircase outline inside. The length of each cut, vertical or horizontal, must be equal to the length of the top right flat edge (the one jutting out). This cutting procedure produces two parts, A and B (Figure A3.7).

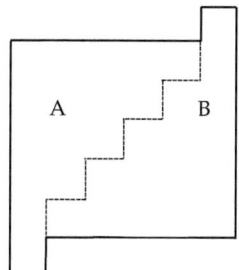

Figure A3.7 Figure with Cut-Outs

By sliding A up or B down, the parts will interlock to produce a rectangle (Figure A3.8).

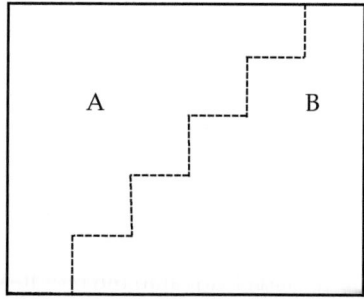

Figure A3.8 Produced Rectangle

5. *Answer*: see solution

Solution

One possible solution is shown in Figure A3.9. There are other ways to cut up the triangle. Note that none of the seven triangles is a right triangle even if barely so.

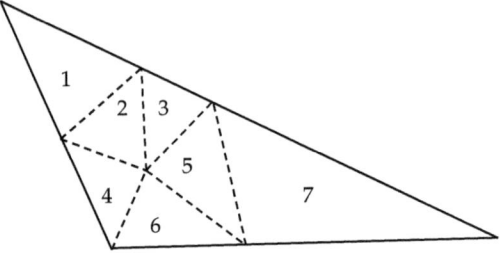

Figure A3.9 Seven Triangles

6. *Answer*: see solution

Solution

Take the second stick in the "VII" denominator (where it represents the numeral for "one" in the Roman system), and join it laterally to the "V," leaving everything else alone (Figure A3.10).

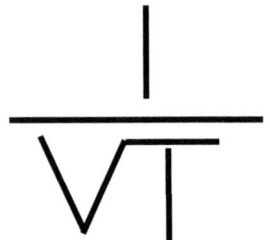

Figure A3.10 New Fraction

The denominator now shows the square root of one, $\sqrt{I}$, which is equal to one, as is the entire fraction, consisting of one, I, over the square root of one: $I/\sqrt{I} = I$ (one).

7. *Answer*: see solution

Solution

Let W stand for any white disc and B for any black disc. Subscripts are used to keep track of each individual disc as it moves to new positions. Below is the sequence of 15 movements required to reverse the initial layout, using the given rules. Note that the sequence can start with a "B" rather than a "W," but the end result would be the same. Remember that the W's move rightward and the B's leftward.

    0. W_1 W_2 W_3 _ B_1 B_2 B_3 → Initial layout
    1. W_1 W_2 _ W_3 B_1 B_2 B_3 → Slide "W_3" into the middle space
    2. W_1 W_2 B_1 W_3 _ B_2 B_3 → Move "B_1" over "W_3" into the empty space
    3. W_1 W_2 B_1 W_3 B_2 _ B_3 → Slide "B_2" into the empty space

4. $W_1\ W_2\ B_1\ \_\ B_2\ W_3\ B_3 \rightarrow$ Move "W_3" over "B_2" into the empty space
5. $W_1\ \_\ B_1\ W_2\ B_2\ W_3\ B_3 \rightarrow$ Move "W_2" over "B_1" into the empty space
6. $\_\ W_1\ B_1\ W_2\ B_2\ W_3\ B_3 \rightarrow$ Slide "W_1" into the empty space
7. $B_1\ W_1\ \_\ W_2\ B_2\ W_3\ B_3 \rightarrow$ Move "B_1" over "W_1" into the empty space
8. $B_1\ W_1\ B_2\ W_2\ \_\ W_3\ B_3 \rightarrow$ Move "B_2" over "W_2" into the empty space
9. $B_1\ W_1\ B_2\ W_2\ B_3\ W_3\ \_ \rightarrow$ Move "B_3" over "W_3" into the empty space
10. $B_1\ W_1\ B_2\ W_2\ B_3\ \_\ W_3 \rightarrow$ Slide "W_3" over into the empty space
11. $B_1\ W_1\ B_2\ \_\ B_3\ W_2\ W_3 \rightarrow$ Move "W_2" over "B_3" into the empty space
12. $B_1\ \_\ B_2\ W_1\ B_3\ W_2\ W_3 \rightarrow$ Move "W_1" over "B_2" into the empty space
13. $B_1\ B_2\ \_\ W_1\ B_3\ W_2\ W_3 \rightarrow$ Slide "B_2" into the empty space
14. $B_1\ B_2\ B_3\ W_1\ \_\ W_2\ W_3 \rightarrow$ Move "B_3" over "W_1" into the empty space
15. $B_1\ B_2\ B_3\ \_\ W_1\ W_2\ W_3 \rightarrow$ Slide "W_1" into the empty space $\rightarrow$ Reverse layout

8. *Answer*: 4 inches

Solution

Sketch the volume layout, marking the given dimensions and locating the covers and pages as they would appear when looking at each volume on the shelf (Figure A3.11).

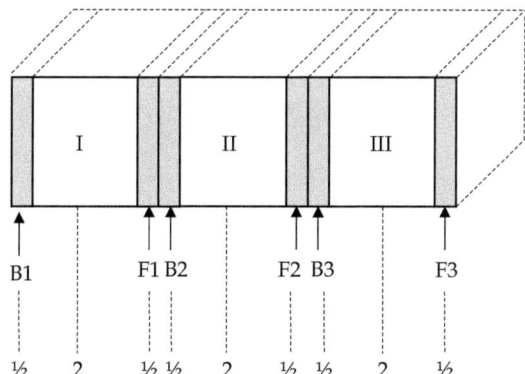

Figure A3.11 Volumes Laid Out on a Shelf

- B1 = back cover of the first volume
- B2 = back cover of the second volume
- B3 = back cover of the third volume
- F1 = front cover of the first volume
- F2 = front cover of the second volume
- F3 = front cover of the third volume

The bookworm started on the first page of Volume I. Where is that on the diagram? If we were to pick up an actual book, labeled Volume I, we would see that it is where it is marked in the diagram below, just before F1 (Figure A3.12). Now, where is the last page of Volume III on the diagram? Again, if we were to pick up an actual book, labeled Volume III, we would see that it is where it is marked below, just after B3 (Figure A3.12).

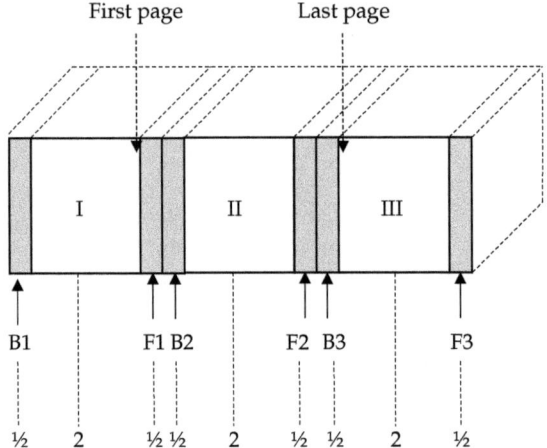

Figure A3.12 First and Last Pages in the Bookworm's Path

Now, let us trace the bookworm's path from the first page of Volume I to the last page of Volume III (Figure A3.13).

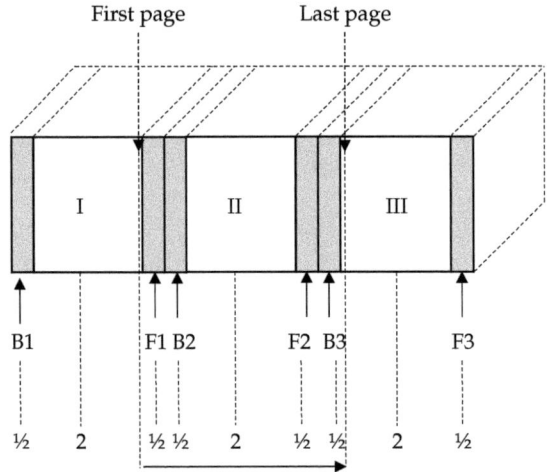

Figure A3.13 Length of Bookworm's Path

As can be seen the total distance covered by the bookworm is:

$$1/2 + 1/2 + 2 + 1/2 + 1/2 = 4$$

9. *Answer*: see solution

Solution

Nothing has happened to the tenth line. Because of the slide, it has become coincident with the side of the reduced sheet:

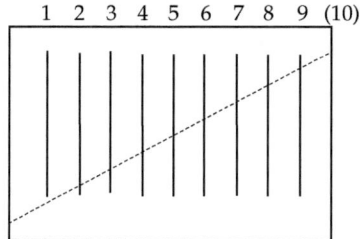

1 2 3 4 5 6 7 8 9 (10)

Figure A3.14 The "Missing" Tenth Line

10. *Answer*: white

Solution

How is it possible that the scientists travel as stipulated and end up back at the camp? On a two-dimensional surface, this is, of course, nonsensical. But the Earth's surface is spherical, not planar. The camp is pitched at the North Pole, and the movements of the scientists, as described by the puzzle, will lead them back to the Pole. Hence, the bear is a polar bear, which is white. The South Pole is excluded because there are no bears there.

Chapter 4

1. *Answer*: 40 centimeters

Solution

Here is the breakdown:

- Let the total length of the fish be x centimeters.
- Half this length is $1/2 \times x = x/2$ centimeters.
- If we add this to 20 centimeters, we get the length of the fish:

$$x = x/2 + 20$$
$$2x = x + 40$$
$$x = 40$$

2. *Answer*: 36

Solution

Here is the breakdown:

- Let the number of sheep be x.
- If we add to that number "again as many as there are in the (original) flock," which is x, there would then be $x + x = 2x$ sheep.
- To this $2x$ "half again of the original number," or $1/2 \times x = x/2$, must be added, which translates as: $x/2 + 2x$, which is simplified to $5x/2$.
- To this $5x/2$, "a quarter of the original number" must be added, or $1/4 \times x = x/4$, which translates as: $x/4 + 5x/2 = 11x/4$.
- Adding "one more sheep" to this amount ($11x/4$) would make 100:

$$11x/4 + 1 = 100$$
$$11x/4 = 100 - 1$$
$$11x/4 = 99$$
$$11x = 396$$
$$x = 36$$

3. *Answer*: 57

Solution

Here is the breakdown:

- Let x stand for the total number of beads in the necklace.
- There are 10 silver beads, 2 blue beads, 1 white bead, and a number of red and turquoise beads on the chain.
- Let y stand for the number of the red beads and z for the number of turquoise beads.
- The total number of beads, x, is made up of the silver (10), blue (2), white (1), red (y), and turquoise (z) beads:

$$x = 10 + 2 + 1 + y + z$$
$$x = y + z + 13$$

- The number of red beads, y, is equal to one-third the total number of beads, $1/3 \times x = x/3$, minus three beads:

$$y = x/3 - 3$$
$$y = (x - 9)/3$$

- The number of turquoise beads (z) is $1/2$ the number of red beads (y), which is $1/2 \times (x - 9)/3$, plus twice the number of silver beads, which is $10 \times 2 = 20$:

$$z = 1/2 \times (x - 9)/3 + 20$$
$$z = (x - 9)/6 + 20$$
$$6z = (x - 9) + 120$$
$$6z = x + 111$$
$$z = (x + 111)/6$$

- Substitute the values of y and z into the equation representing the total number of beads (above), $x = y + z + 13$, to solve for x:

$$x = y + z + 13$$
$$x = (x - 9)/3 + (x + 111)/6 + 13$$
$$x = (2x - 18)/6 + (x + 111)/6 + 13$$
$$6x = (2x - 18) + (x + 111) + 78$$
$$6x = 3x + 171$$
$$3x = 171$$
$$x = 57$$

There were 57 beads in total on the necklace, broken down as follows:

Silver: 10
Blue: 2
White: 1
Red: 16
Turquoise: 28

4. *Answer*: 16.625

Solution

Then problem is solved easily by setting up an equation, using modern notation:

- Let x stand for the unknown quantity.
- 1/7th of this quantity is $x/7$.
- Adding these together produces the sum of 19:

$$x + x/7 = 19$$
$$7x + x = 133$$
$$8x = 133$$
$$x = 16.625$$

5. *Answer*: 577 and 7/9 staters; 422 and 2/9 staters

Solution

Using modern algebraic notation, the breakdown is as follows:

- Let x stand for the amount of staters to be given to the older heir.
- From this, it follows that $(1000 - x)$ is what will be given to the younger heir, since the father had 1000 staters to give out to the two heirs, and we know that x is what the older one will receive, which means that $(1000 - x)$ is what is left over for the younger one.
- The "fifth part of the older one's share" translates to $1/5 \times x = x/5$.
- The "fourth part of what falls to the younger one" translates to $1/4 \times (1000 - x) = (1000 - x)/4$.
- The difference between these two amounts is 10 staters. This is the required equation:

$$x/5 - (1000 - x)/4 = 10$$
$$4x - 5(1000 - x) = 200$$
$$4x - 5000 + 5x = 200$$
$$9x + 5x = 200 + 5000$$
$$9x = 5200$$
$$x = 577 \text{ and } 7/9 \text{ staters}$$

- This is what the older child will receive; so the younger child will get:

$$1000 - x = 1000 - 577 \text{ and } 7/9, \text{ or } 422 \text{ and } 2/9 \text{ staters.}$$

6. *Answer*: The woman is 40 years of age and her friend is 20 years of age.

Solution

The breakdown is as follows:

- Let x and y be the respective ages of the woman and her friend.
- Ten years ago, the woman was $(x - 10)$ years old and the friend $(y - 10)$ years old.
- At that time, the friend's age, $(y - 10)$, was 1/3 the woman's age, $1/3 \times (x - 10) = (x - 10)/3$:

$$(y - 10) = (x - 10)/3$$
$$3(y - 10) = (x - 10)$$
$$3y - 30 = x - 10$$
$$3y - x = 20$$

- Ten years from the time of the conversation, the woman will be $(x + 10)$ years old, and her friend $(y + 10)$ years old.
- The woman will be 20 years older at that future time, which means that the difference of their ages will be 20:

$$(x + 10) - (y + 10) = 20$$
$$x - y = 20$$
$$x = y + 20$$

- We can now substitute this value of x into the equation above, $(3y - x = 20)$:

$$3y - x = 20$$
$$3y - (y + 20) = 20$$
$$2y - 20 = 20$$
$$2y = 40$$
$$y = 20 \ (= \text{ the friend's age})$$

- Since the woman is twice as old, she is 40. Note that there are other ways to solve this problem.

7. *Answer*: 1 hour and 12 minutes

Solution

Here is the breakdown:

- Angela can complete the job in 3 hours. So, on average, she completes 1/3 of the job in 1 hour (and of course, 2/3 in 2 hours and 3/3, or the whole job, in 3 hours).
- Bernice can complete the job in 2 hours. So, on average, she completes 1/2 of the job per hour.
- We are asked to determine how much time it would take when the two are working together.
- Let us represent that time with x.

- The two of them working together will complete the job in x hours. So, in 1 hour they would complete $1/x$ of the job (on average).
- Angela would complete $1/3$ of the job and Bernice $1/2$ of the job working alone in 1 hour.
- Summing their two rates, we get:

$$1/3 + 1/2 = 1/x$$
$$5/6 = 1/x$$
$$5x = 6$$
$$x = 6/5$$

- This means that it will take Angela and Bernice $6/5$ hours, or 1 hour and 12 minutes, to complete the job working together.

8. *Answer*: 60

Solution

Here is the breakdown:

- Let x be Demochares' age.
- He lived "a fourth of his life as a boy," that is, $1/4 \times x = x/4$.
- He lived "a fifth as a youth," that is, $1/5 \times x = x/5$.
- He lived "a third as a man," that is, $1/3 \times x = x/3$.
- He lived 13 years "in his old age."
- Adding all these up, we get Demochares' age:

$$x = x/4 + x/5 + x/3 + 13$$
$$x = 47x/60 + 13$$
$$60x = 47x + 780$$
$$13x = 780$$
$$x = 60$$

9. *Answer*: One horse will have traveled 225 feet and the other one 200 feet when they meet.

Solution

Here is the breakdown:

- There are 425 feet between the two horses. Let us call them A and B.
- Let the distance that the first horse, A, traveled at the point of meeting the other horse be x miles. Then the distance traveled by the other horse, B, will be $(425 - x)$.
- Distance equals rate multiplied by time ($D = R \times T$).
- The horses used the same amount of time to get to the meeting point. And we are told the rate of each horse. Therefore:

Horse A
$D = x$
$R = 45$ feet per minute
$T = t$ (constant)
$D = R \times T$
$x = 45 \times t$
$t = x/45$

Horse B
D = (425–x)
R = 40 feet per minute
T = t (constant)
D = R × T
(425 – x) = 40 × t
t = (425 – x)/40

- Since *t* is common to both, we can set up the following equivalence:

x/45 = (425 – x)/40
40x = 45 (425 – x)
40x = 19125 – 45x
85x = 19125
x = 225 (distance covered by the first horse)
425 – x = 425 – 225 = 200 (distance covered by the second horse)

10. *Answer*: $20.24

Solution

Here is the breakdown:

- Let *x* represent the number of pounds of potatoes that the farmer bought at 92¢ per pound.
- This means that he spent $0.92x for the potatoes.
- If the price had been 4¢ per pound less, that is 88¢ per pound, the farmer could have bought one more pound, or (x + 1) pounds, for the same money.
- This means that he would have spent $0.88 (x + 1). The total cost would have been the same. Therefore:

$$0.92x = 0.88 (x + 1)$$
$$92x = 88x + 88$$
$$4x = 88$$
$$x = 22$$

- So, the farmer bought 22 pounds of potatoes. At 92¢ per pound, he spent $20.24.

Chapter 5

1. *Answer*: 23 cents

Solution

Here is the breakdown:

- Let *x* stand for the price of a button in cents.
- The daughter says that she bought two buttons less than the price of a button, namely (x – 2) buttons.
- The total cost of the (x – 2) buttons was, therefore, x (x – 2), the number of buttons times the cost per button.

- In total, she spent $4.83, which in cents is 483. Therefore:

$$x(x-2) = 483$$
$$x^2 - 2x = 483$$
$$x^2 - 2x - 483 = 0$$

- Using the factorization method:

$$(x-23)(x+21) = 0$$

- So, either factor can be equal to zero:

$$(x-23) = 0 \rightarrow x = 23$$
$$(x+21) = 0 \rightarrow x = -21$$

- We discard $x = -21$, since the price could not be a negative value.
- So, $x = 23$, which means that the price of a button was 23 cents.

2. *Answer*: 64

Solution

Here is the breakdown:

- Let x be the number of sons the grandmother has.
- Each son will have $(x-1)$ brothers.
- Each son has, himself, as many sons as he has brothers, that is, each son will have $(x-1)$ sons of his own.
- Since there are x sons, there are $x(x-1)$ grandchildren in total, or $(x^2 - x)$.
- The sum of the number of sons and grandchildren is thus:

$$x + (x^2 - x) = x^2$$

- This stands for the grandmother's age, since, as she points out, "The combined number of my sons and grandsons equals my age."
- Now, x^2 stands for a square number between 50 and 70; and the only such number in this range is 64 (which is 8^2).
- So the grandmother is 64 years old.

3. *Answer*: 25

Solution

Here is the breakdown:

- Let x represent the number of coconuts in the original pile.

   First sailor

- He gives one coconut to the monkey, leaving $(x-1)$ coconuts in the pile.
- From this pile he takes 1/3:

$$(x-1)/3$$

- After doing so, this leaves the following number of coconuts behind—the number of coconuts left, $(x - 1)$, minus the number taken by the sailor, $(x-1)/3$:

$$(x-1) - (x-1)/3 = (2x-2)/3$$

Second sailor

- He gives one coconut to the monkey from the pile left by the previous sailor, represented by $(2x - 2)/3$ above, leaving the following number of coconuts:

$$(2x-2)/3 - 1 = (2x-5)/3$$

- From this pile, he takes 1/3:

$$1/3 \times (2x-5)/3 = (2x-5)/9$$

- After doing so, the number of coconuts left over is determined by subtracting what the second sailor took, $(2x - 5)/9$, from the number in the pile above, $(2x - 5)/3$:

$$(2x-5)/3 - (2x-5)/9 = (4x-10)/9$$

Third sailor

- He gives yet another coconut to the monkey, taken from the number left by the second sailor, which is $(4x - 10)/9$:

$$(4x-10)/9 - 1 = (4x-19)/9$$

- He takes 1/3 of this number:

$$1/3 \times (4x-19)/9 = (4x-19)/27$$

- After doing so, the number of coconuts left over is determined by subtracting what the third sailor took, $(4x - 19)/27$, from the number in the pile above, $(4x - 19)/9$:

$$(4x-19)/9 - (4x-19)/27 = (8x-38)/27$$

- Now, this means that there are $(8x - 38)/27$ coconuts left in the morning.

- We are told that this number is less than 10 and that it is a number that can be evenly divided among the three sailors.
- This is where Diophantine-thinking comes into play: $(8x - 38)/27$ must have an integral value, which means that the expression must be equal to a multiple of 3 that is less than 10. The possibilities are:

$$(8x - 38)/27 = 3$$
$$(8x - 38)/27 = 6$$
$$(8x - 38)/27 = 9$$

- The only integral value emerges when $x = 25$ in the second equation above:

$$(8x - 38)/27 = 6$$
$$8x - 38 = 162$$
$$8x = 200$$
$$x = 25$$

- So, there were 25 coconuts in the original pile.

4. *Answer*: 20

Solution

Here is the breakdown:

- Let the number to be added be n.
- Then $(100 + n)$ and $(20 + n)$ represent the numbers that result when n is added respectively to 100 and to 20.
- We are told that $(100 + n)$ is to $(20 + n)$ as 3 is to 1, which translates into the following equation:

$$(100 + n)/(20 + n) = 3/1$$
$$(100 + n) = 3 (20 + n)$$
$$100 + n = 60 + 3n$$
$$2n = 40$$
$$n = 20 \text{ (the required number)}$$

5. *Answer*: 3, 4, and 5

Solution

This is, actually, the Pythagorean theorem in disguise, satisfied by the first triplet of numbers, namely {3, 4, 5}. Here is the breakdown:

- If the first integer is represented by n, then the integer right after is $(n + 1)$, and the one after that is $(n + 2)$.
- The problem states that the sum of the squares of the first two integers equals the square of the third integer:

$$n^2 + (n + 1)^2 = (n + 2)^2$$
$$n^2 + n^2 + 2n + 1 = n^2 + 4n + 4$$
$$2n^2 + 2n + 1 = n^2 + 4n + 4$$
$$n^2 - 2n + 1 = 4$$
$$n^2 - 2n - 3 = 0$$

Factoring :

$$(n - 3)(n + 1) = 0$$
$$n = \{3, -1\}$$

- We discard the negative root because we are looking only for positive numbers, Thus: $n = 3$.
- The consecutive numbers are thus 3, 4, and 5.

6. *Answer*: 8, 15, and 17

Solution

This is also the Pythagorean theorem in disguise, satisfied this time by another triplet, namely {8, 15, 17}. Here is the breakdown:

- Let the smallest integer be n.
- The largest integer, which is greater by 9 than the smallest one, is thus $(n + 9)$.
- The integer that is greater by 7 than the smallest one is $(n + 7)$.
- The problem states that the square of the largest integer is equal to the sum of the squares of the other two integers:

$$(n + 9)^2 = n^2 + (n + 7)^2$$
$$n^2 + 18n + 81 = n^2 + n^2 + 14n + 49$$
$$n^2 + 18n + 81 = 2n^2 + 14n + 49$$
$$n^2 - 4n - 32 = 0$$

Factoring :

$$(n - 8)(n + 4) = 0$$
$$n = \{8, -4\}$$

- We discard the negative root because we are looking only for positive integers. Thus: $n = 8$.
- The other two integers are:

$$(n + 7) = 8 + 7 = 15$$
$$(n + 9) = 8 + 9 = 17$$

7. *Answer*: 24 rows containing 32 bushes each

Solution

Here is the breakdown:

- Let the total number of rows be represented by x.
- In each row, there will be eight more bushes than the number of rows, or $(x + 8)$ bushes per row.
- The number of rows × number of bushes per row = total number of bushes in the field, which is equal to 768:

$$x\,(x + 8) = 768$$
$$x^2 + 8x - 768 = 0$$

Factoring :

$$(x + 32)(x - 24) = 0$$

So :
$$x = \{-32, +24\}$$

- We discard the minus root, since it has no meaning in this case.
- Thus, $x = 24$. This is the number of rows. The number of bushes per row is 8 more than that, or 32.
- Therefore, the farmer will have an arrangement of 24 rows containing 32 bushes each.

8. *Answer*: 30

Solution

Here is the breakdown:

- Let the merchant's age be x.
- Adding 5 to this translates as $(x + 5)$.
- Multiplying this by the merchant's age translates as $x\,(x + 5)$.
- This equals 1050:

$$x\,(x + 5) = 1050$$
$$x^2 + 5x - 1050 = 0$$

Factoring :

$$(x + 35)(x - 30) = 0$$
So :
$$x = \{30, -35\}$$

- We discard the minus root, since it has no meaning in this case.
- Therefore, the merchant was 30 years old.

9. *Answer*: 1480 pounds

Solution

Here is the breakdown:

- Let the original sum of money be x.

- "During the first year he spent 100 pounds," translates as:

$$(x - 100) \ (= \text{the remaining sum})$$

- To this, the merchant "added one third of it," which can be rephrased as follows. One-third of the remaining sum is $(x - 100)/3$, and to this the merchant added the sum of $(x - 100)$:

$$(x - 100) + (x - 100)/3 = (4x - 400)/3 \ (= \text{ remaining sum})$$

- "During the next year he again spent 100 pounds," means that he took away 100 from the remaining sum above:

$$(4x - 400)/3 - 100 = (4x - 700)/3 \ (= \text{ remaining sum})$$

- The merchant then "increased the remaining sum by one third of it," which can be rephrased as follows. To the remaining sum above, $(4x - 700)/3$, the merchant added 1/3 of it, or $1/3 \times (4x - 700)/3$, simplified to $(4x - 700)/9$:

$$(4x - 700)/3 + (4x - 700)/9 = (16x - 2800)/9 \ (= \text{ remaining sum })$$

- "During the third year he again spent 100 pounds," means that he took away 100 from the remaining sum above:

$$(16x - 2800)/9 - 100 = (16x - 3700)/9 \ (= \text{ remaining sum})$$

- The merchant then added the remaining sum above, $(16x - 3700)/9$, to 1/3 of the remaining sum, which is $1/3 \times (16x - 3700)/9$, simplified to $(16x - 3700)/27$:

$$(16x - 3700)/9 + (16x - 3700)/27 = (64x - 14800)/27$$

- The remaining sum above is equal to twice the original sum, or $2x$:

$$2x = (64x - 14800)/27$$
$$54x = 64x - 14800$$
$$10x = 14800$$
$$x = 1480$$

- The original sum was, therefore, 1480 pounds.

10. *Answer*: 5

Solution

Here is the breakdown:

- Let the number be n.
- The two successive integers after it are $(n + 1)$ and $(n + 2)$.
- Eight times the number, n, plus 2 is $(8n + 2)$.
- This is equal to the product of the two successive integers:

$$(n + 1)(n + 2) = (8n + 2)$$
$$n^2 + 3n + 2 = 8n + 2$$
$$n^2 - 5n + 2 = 2$$
$$n^2 - 5n = 0$$

Factoring :

$$n\,(n - 5) = 0$$
$$n = \{0, 5\}$$

- We discard 0, since it has no meaning here. Therefore, $n = 5$.

Chapter 6

1. *Answer*: The gentleman is the lady's uncle.

Solution

Only siblings have the same (biological) mother. In this case, the lady is talking about a brother and sister, with the sister being her own mother. The brother, therefore, is her uncle.

2. *Answer*: the boy's grandmother

The father of the young man married the painter's daughter. He is thus the son-in-law of the painter. Now, his own son is the painter's grandson, which means that the painter is his grandmother.

3. *Answer*: The boy is the man's son.

Solution

When the man alludes to "my father's son" he can only be talking about himself, since he is an only child. So, he himself is "that boy's father," which makes the boy his son.

4. *Answer*: the woman herself

Solution

Since the woman is an only child, she is the only daughter of her father. So, if the father of the woman in the portrait is her father as well, as she tells us, then she is looking at herself.

5. *Answer*: She is the young man's mother.

Solution

Since the young man is an only child, then when he refers to his "mother's only child" he is speaking about himself. So, she is his mother, given that her son, or as he puts it, "that woman's son," is himself ("my mother's only child").

6. *Answer*: not possible

Solution

The trap in this puzzle is the meaning of the word *widow*. A man who leaves a *widow* is a dead man.

7. *Answer*: They are cousins.

Solution

The son of the first woman's mother is her brother. Since the brother is the second woman's cousin, then the first woman is her cousin as well.

8. *Answer*: Yes

Solution

She can if the woman is her mother, meaning that the woman's husband is her father.

9. *Answer*: see solution

Solution

- Let us call the three mothers A, B, and C.
- A is B's mother.
- B is the mother of two of the children, D and E.
- D and E are thus a brother and a sister.
- C is A's sister and thus her sibling, making three siblings in total.
- C is the mother of the third child, F.
- The three children are cousins.
- B is the aunt of F, and C the aunt of D and E.
- The actual six people were thus: A, B, C, D, E, and F.

10. *Answer*: The computer salesclerk was out the $3 device and $7 from his pocket—$10 in total.

Solution

This is similar to the one discussed in the *Recreational Logic* section of this chapter. Like that puzzle, the deception is to be found in the layout of the numerical facts. Here's how they can be laid out in order to avoid the apparent discrepancy.

- First, the computer salesclerk received nothing for the $3 device, since the counterfeit $10 bill was worth nothing. So, from the outset, he was out $3 (the price of the device). That $3 went to the customer.
- Now, let us consider what happened in the other transaction—the one between the computer and clothing salesclerks. The computer salesclerk received 10 genuine $1 bills for the fake $10 bill. So, the clothing salesclerk was out $10 at this point.
- When the computer salesclerk got back to his store, he gave $7 of the 10 good bills to the customer and put the remaining $3 in his pocket. The end result of this transaction was that the computer salesclerk was out another $7, while the customer gained another $7. Altogether, the customer gained $10—a $3 device and $7 in good bills. That ends the computer salesclerk's transaction with the customer.

- At this point, the computer salesclerk was out the $3 for the device, not the $7 that he gave back as change to the customer—that came out of the pocket of the clothing salesclerk. When the clothing salesclerk asked for her $10 back, the computer salesclerk still had the $3 in his pocket left over from the $10 she had given him previously—remember that the other $7 went to the customer.
- So, he gave her back her $3 and made up the $7 difference from his own pocket. In total, therefore, the computer salesclerk was out the $3 device and the $7 from his pocket—$10 in total.

Chapter 7

1. *Answer*: three trips are required

Solution

Below is a graphic solution: M = man, W = woman, C = cat (Figure A7.1).

Original Side	Boat	Other Side
0. M W C	– –	– – –
1. M _ _	W C →	– – –
2. M _	← _ W	_ _ C
3. _ _ _	M W →	_ _ C
0. _ _ _	– –	M W C

Figure A7.1 Graphic Model for Exploration Problem 1

2. *Answer*: seven trips are required

Solution

There are several ways that the trips over can be carried out, but the end result will be the same—seven trips will be required. Below is one possibility: M_1 = one man, M_2 = the other man, W_1 = one woman, W_2 = the other woman, C = the cat (Figure A7.2).

Original Side	Boat	Other Side
0. M_1 M_2 W_1 W_2 C	– –	– – – – –
1. M_1 M_2 _ W_2 _	W_1 C →	– – – – –
2. M_1 M_2 _ W_2 _	← W_1 _	– – – – C
3. M_1 M_2 _ _ _	W_1 W_2 →	– – – – C
4. M_1 M_2 _ _ _	← W_1 _	– – – W_2 C
5. M_1 _ _ _ _	W_1 M_2 →	– – – W_2 C
6. M_1 _ _ _ _	← M_2 _	– – W_1 W_2 C
7. _ _ _ _ _	M_1 M_2 →	– – W_1 W_2 C
0. _ _ _ _ _	– –	M_1 M_2 W_1 W_2 C

Figure A7.2 Graphic Model for Exploration Problem 2

3. *Answer*: seven trips are required

Solution

There are several ways to plan the trips back and forth, but the end result will be the same—seven trips in total are required. Below is one possibility: C_1 = one child, C_2 = the other child, W = the woman, C = the cat, D = the dog (Figure A7.3).

Original Side	Boat	Other Side
0. C_1 C_2 W C D	$- -$	$- - - - -$
1. $\_$ C_2 $\_$ C D	C_1 W$\rightarrow$	$- - - - -$
2. $\_$ C_2 $\_$ C D	$\leftarrow$ W $\_$	C_1 $\_ \_ \_ \_$
3. $\_$ C_2 $\_ \_$ D	W C$\rightarrow$	C_1 $\_ \_ \_ \_$
4. $\_$ C_2 $\_ \_$ D	$\leftarrow$ W $\_$	C_1 $\_ \_$ C $\_$
5. $\_$ C_2 $\_ \_ \_$	W D$\rightarrow$	C_1 $\_ \_$ C $\_$
6. $\_$ C_2 $\_ \_ \_$	$\leftarrow$ W	C_1 $\_ \_$ C D
7. $\_ \_ \_ \_ \_$	W $C_2$$\rightarrow$	C_1 $\_ \_$ C D
0. $\_ \_ \_ \_ \_$	$- -$	C_1 C_2 W C D

Figure A7.3 Graphic Model for Exploration Problem 3

4. *Answer*: seven trips

Solution

There are several ways to plan the trips back and forth, but the end result will be the same—seven trips in total. Below is one possibility: T_1 = one traveler, T_2 = the other traveler, W = wolf, G = goat, C = cabbage (Figure A7.4).

Original Side	Boat	Other Side
0. T_1 T_2 W G C	$- -$	$- - - - -$
1. $\_$ T_2 W $\_$ C	T_1 G $\rightarrow$	$- - - - -$
2. $\_$ T_2 W $\_$ C	$\leftarrow T_1$ $\_$	$\_ \_ \_$ G $\_$
3. $\_ \_$ W $\_$ C	T_1 T_2 $\rightarrow$	$\_ \_ \_$ G $\_$
4. $\_ \_$ W $\_$ C	$\leftarrow$ T_1 $\_$	$\_$ T_2 $\_$ G $\_$
5. $\_ \_$ W $\_ \_$	T_1 C $\rightarrow$	$\_$ T_2 $\_$ G $\_$
6. $\_ \_$ W $\_ \_$	$\leftarrow$ T_1 $\_$	$\_$ T_2 $\_$ G C
7. $\_ \_ \_ \_ \_$	T_1 W $\rightarrow$	$\_$ T_2 $\_$ G C
0. $\_ \_ \_ \_ \_$	$- -$	T_1 T_2 W G C

Figure A7.4 Graphic Model for Exploration Problem 4

5. *Answer*: five trips

Solution

There are other ways to plan the trips back and forth, but the end result will be the same—five trips in total. Below is one possibility: B = brother, S = sister, C_1 = one cousin, C_2 = the other cousin (Figure A7.5).

Original Side	Boat	Other Side
0. $B S C_1 C_2$	--	----
1. $B \_ C_1 \_$	$S C_2 \rightarrow$	----
2. $B \_ C_1 \_$	$\leftarrow S \_$	$--- C_2$
3. $\_\_ C_1 \_$	$B S \rightarrow$	$--- C_2$
4. $\_\_ C_1 \_$	$\leftarrow S$	$B \_\_ C_2$
5. $\_\_\_\_$	$S C_1 \rightarrow$	$B \_\_ C_2$
0. $\_\_\_\_$	--	$B S C_1 C_2$

Figure A7.5 Graphic Model for Exploration Problem 5

6. *Answer*: No

Solution

At some point the wolf, goat, or cabbage would be left alone, and the result would be disastrous.

7. *Answer*: 15 trips

Solution

This is similar to Loyd's problem, and thus needs an island as a transit stop: H_1 and W_1 = first husband and wife pair; H_2 and W_2 = second husband and wife pair; H_3 and W_3 = third husband and wife pair; H_4 and W_4 = fourth husband and wife pair (Figure A7.6).

Original Side	Boat	Island	Boat	Other Side
0. $H_1 W_1 H_2 W_2 H_3 W_3 H_4 W_4$	--	--	--	--------
1. $H_1 \_ H_2 \_ H_3 W_3 H_4 W_4$	$W_1 W_2 \rightarrow$	$\_\_$ (bypass)	$W_1 W_2 \rightarrow$	----------
2. $H_1 \_ H_2 \_ H_3 W_3 H_4 W_4$	$\leftarrow W_2 \_$	$\_\_$ (bypass)	$\leftarrow W_2 \_$	$\_ W_1 --------$
3. $H_1 \_ H_2 \_ H_3 \_ H_4 W_4$	$W_2 W_3 \rightarrow$	$\_\_$ (stop)	--	$\_ W_1 ------$
4. $H_1 \_ H_2 \_ H_3 \_ H_4 W_4$	$\leftarrow W_2 \_$	$W_3 \_$ (double back)	--	$\_ W_1 ------$
5. $H_1 \_\_\_\_ H_3 \_ H_4 W_4$	$H_2 W_2 \rightarrow$	$W_3 \_$ (bypass)	$H_2 W_2 \rightarrow$	$\_ W_1 ------$
6. $H_1 \_\_\_\_ H_3 \_ H_4 W_4$	$\leftarrow W_1 \_$	$W_3 \_$ (bypass)	$\leftarrow W_1 \_$	$\_\_ H_2 W_2 ------$
7. $\_\_\_\_ H_3 \_ H_4 W_4$	$H_1 W_1$	$W_3 \_$ (bypass)	$H_1 W_1 \rightarrow$	$\_\_ H_2 W_2 ------$
8. $\_\_\_\_ H_3 \_ H_4 W_4$	$\leftarrow W_3 \_$	$W_1 \_$ (change)	$\leftarrow W_1 \_$	$H_1 \_ H_2 W_2 ----$
9. $\_\_\_\_ H_4 W_4$	$H_3 W_3 \rightarrow$	$W_1 \_$ (bypass)	$H_3 W_3 \rightarrow$	$H_1 \_ H_2 W_2 ----$
10. $\_\_\_\_ H_4 W_4$	$\leftarrow W_3 \_$	$W_1 \_$ (bypass)	$\leftarrow W_3 \_$	$H_1 \_ H_2 W_2 H_3 ----$
11. $\_\_\_\_ H_4 \_$	$W_3 W_4 \rightarrow$	$W_1 \_$ (bypass)	$W_3 W_4 \rightarrow$	$H_1 \_ H_2 W_2 H_3 ----$
12. $\_\_\_\_ H_4 \_$	$\leftarrow W_4 \_$	$W_1 \_$ (bypass)	$\leftarrow W_4 \_$	$H_1 \_ H_2 W_2 H_3 W_3 \_\_$
13. $_____$	$H_4 W_4 \rightarrow$	$W_1 \_$ (bypass)	$H_4 W_4 \rightarrow$	$H_1 \_ H_2 W_2 H_3 W_3 \_\_$
14. $_____$	--	$W_1 \_$	$\leftarrow H_1 \_$	$\_\_ H_2 W_2 H_3 W_3 H_4 W_4$
15. $_____$	--	$\_\_$ (pick up)	$H_1 W_1 \rightarrow$	$\_\_ H_2 W_2 H_3 W_3 H_4 W_4$
0. $_____$	--	--	--	$H_1 W_1 H_2 W_2 H_3 W_3 H_4 W_4$

Figure A7.6 Graphic Model for Exploration Problem 7

8. *Answer*: 12

Solution

Here is the breakdown:

- There are three different ways to go from A to B.
- From B, there are four different ways to get to C.

- So, for each route taken from A to get to B, there are four routes that can then be taken to get to C.
- Therefore, there are 3 × 4 = 12 different ways altogether to get from A to C.

9. *Answer*: 380 possible outcomes. If only two specific members are to be elected to the presidency, there are only 38 possible outcomes to the election.

Solution

Here is the breakdown:

- Once a president is chosen from among 20 members, there are 19 members left to be chosen as vice-president.
- So, there are 20 × 19 = 380 possible ways to elect a president and vice-president.
- If only Brenda and Heather can be selected for the presidency, the choice for that position is restricted to: 2 × 1 = 2.
- Now, for each one of these, there are 19 members left (including the unelected Brenda or Heather) for the vice-president position.
- So, in this case, the number of possible election pairs is 2 × 19 = 38.

10. *Answer*: 792 kinds of soup

Solution

Here is the breakdown:

- The chef has 12 kinds of vegetables at her disposal.
- However, she is restricting the number to exactly five different kinds.
- So, for each of the 12 vegetables, she can choose 11 of those that remain, and then 10 of those that remain, after which she can choose 9 of those that remain, and finally 8 of the remainder.
- So, in total, she has 12 × 11 × 10 × 9 × 8 = 95,040 possible choices for making the soup.
- The order in which these are chosen is irrelevant.
- How many redundant choices are there among these? There are 5! = 5 × 4 × 3 × 2 × 1 = 120 of them.
- So, she can make 95,040 ÷ 120 = 792 different kinds of soup.

Chapter 8

1. *Answer*: 128

Solution

Each number increases by a successive power of 2, that is, by 2^n, where $n = \{0, 1, 2, 3, 4, 5, 6, \ldots\}$:
$\{1, 2, 4, 8, 16, 32, 64, \ldots\} = \{2^0, 2^1, 2^2, 2^3, 2^4, 2^5, 2^6, \ldots\}$
So, the next number will be $2^7 = 128$.

2. *Answer*: 2187

Solution

Each number increases by a successive power of 3, that is, by 3^n, where $n = \{0, 1, 2, 3, 4, 5, 6, \ldots\}$:

$\{1, 3, 9, 27, 81, 243, 729, \ldots\} = \{3^0, 3^1, 3^2, 3^3, 3^4, 3^5, 3^6, \ldots\}$
So, the next number will be $3^7 = 2187$.

3. *Answer*: 19

Solution

The sequence consists of the first prime numbers in order.

4. *Answer*: 0.0078125

Solution

This is the decimal version of a series in which each term is a successive power of $1/2^n$, where $n = \{0, 1, 2, 3, 4, 5, 6, \ldots\}$:
$\{1/2^0, 1/2^1, 1/2^2, 1/2^3, 1/2^4, 1/2^5, 1/2^6, \ldots\} = \{1, 0.5, 0.25, 0.125, 0.0625, 0.03125, 0.015625, \ldots\}$
 So, the next number will be: $1/2^7 = 1/128 = 0.0078125$.

5. *Answer*: Each term is the product of all the previous terms, plus one.

Solution

$\{2, 3, 7, 43, 1807, 3{,}263{,}443, \ldots\}$—note that "1" is included in the first-term product:

First Term:	$2 = 2 \times 1$
Second term:	$3 = (2 \times 1) + 1$
Third term:	$7 = (2 \times 1) \times 3 + 1$
Fourth term:	$43 = (2 \times 1 \times 3) \times 7 + 1$

...

[The sequence can be used to prove that there are infinitely many prime numbers, since it can be seen that no terms in the sequence have any prime factors in common.]

6. *Answer*: see solution

Solution

The numbers in the new sequence are the squares of the Fibonacci numbers:
$\{1^2, 1^2, 2^2, 3^2, 5^2, 8^2, 13^2, 21^2, 34^2, 55^2, \ldots\} = \{1, 1, 4, 9, 25, 64, 169, 441, 1156, 3025, \ldots\}$

7. *Answer*: 128

Solution

There are two alternating series embedded in the overall sequence, namely a series, 2^n, and an alternating series, 3^n.

$\{2, 3, 4, 9, 8, 27, 16, 81, 32, 243, 64, 729, \ldots\} = \{2^1, 3^1, 2^2, 3^2, 2^3, 3^3, 2^4, 3^4, 2^5, 3^5, 2^6, 3^6, \ldots\}$

So, the next term in the sequence is $2^7 = 128$.

8. *Answer*: 129

Solution

The general term is $(2^n + 1)$.

First Term ($n = 0$)	→	$(2^n + 1) = (2^0 + 1) = (1 + 1) = 2$
Second Term ($n = 1$)	→	$(2^n + 1) = (2^1 + 1) = (2 + 1) = 3$
Third Term ($n = 2$)	→	$(2^n + 1) = (2^2 + 1) = (4 + 1) = 5$
Fourth Term ($n = 3$)	→	$(2^n + 1) = (2^3 + 1) = (8 + 1) = 9$
Fifth Term ($n = 4$)	→	$(2^n + 1) = (2^4 + 1) = (16 + 1) = 17$
Sixth Term ($n = 5$):	→	$(2^n + 1) = (2^5 + 1) = (32 + 1) = 33$
Seventh Term ($n = 6$):	→	$(2^n + 1) = (2^6 + 1) = (64 + 1) = 65$

...

So, the next term in the series is the eighth one, where $n = 7$:

Eighth Term ($n = 7$): → $(2^n + 1) = (2^7 + 1) = (128 + 1) = 129$

9. *Answer*: 127

Solution

The general term is $(2^n - 1)$.

First Term($n = 0$)	→	$(2^n - 1) = (2^0 - 1) = (1 - 1) = 0$
Second Term($n = 1$)	→	$(2^n - 1) = (2^1 - 1) = (2 - 1) = 1$
Third Term($n = 2$)	→	$(2^n - 1) = (2^2 - 1) = (4 - 1) = 3$
Fourth Term($n = 3$)	→	$(2^n - 1) = (2^3 - 1) = (8 - 1) = 7$
Fifth Term($n = 4$)	→	$(2^n - 1) = (2^4 - 1) = (16 - 1) = 15$
Sixth Term($n = 5$):	→	$(2^n - 1) = (2^5 - 1) = (32 - 1) = 31$
Seventh Term($n = 6$):	→	$(2^n - 1) = (2^6 - 1) = (64 - 1) = 63$

...

So, the next term in the series is the eighth one, where $n = 7$:

Eighth Term ($n = 7$) : → $(2^n - 1) = (2^7 - 1) = (128 - 1) = 127$

10. *Answer*: $L_n = L_{n-1} + L_{n-2}$, with $L_0 = 2$ and $L_1 = 1$ (L = Lucas number)

Solution

The rule is the same one as for the Fibonacci Sequence, except that Lucas starts with $L_0 = 2$ and $L_1 = 1$. Here is a comparison:

F_n	0	1	1	2	3	5	8	13	21	34	55	...
L_n	2	1	3	4	7	11	18	29	47	76	123	...

Chapter 9

1. *Answer*: 100

Solution

Here is the breakdown:

- Without the two furrows the area of the field would be $50 \times 40 = 2000$ (square feet).
- Each furrow is 40 feet long, since it starts at the 50-feet top of the field and goes down the length of its width (40 feet) to the other end.
- The width of each furrow is given as 5 feet. So, the area of one furrow is: $5 \times 40 = 200$ (square feet). The area of two furrows is thus: $200 \times 2 = 400$ (square feet).
- Taking this away from the area of the field leaves $2000 - 400 = 1600$ (square feet) that can be covered by the plots.
- Each plot covers $4 \times 4 = 16$ (square feet). So, the number of plots that can be planted into the field is: $1600 \div 16 = 100$.

2. *Answer*: 3 cups: 1 cup of soda, 1 cup of water, and 1 cup of milk.

Solution

Here is the breakdown:

- Let the actual number of cups filled with soda be s, the actual number of cups filled with water be w, and the actual number of cups filled with milk be m.
- So, the total number of cups is $s + w + m$.
- The actual number of cups filled with soda, equals the total number of cups minus two:

$$s = s + w + m - 2$$
$$w + m = 2$$

- The actual number of cups filled with water also equals the total number of cups minus two:

$$w = s + w + m - 2$$
$$s + m = 2$$

- And the actual number of cups filled with milk also equals the total number of cups minus two:

$$m = s + w + m - 2$$
$$s + w = 2$$

- The three equations are repeated below:

$$w + m = 2$$
$$s + m = 2$$
$$s + w = 2$$

- Since all three equations equal 2, we can equate any of them to any other. So, let us take the first two:

$$w + m = s + m$$
$$w = s$$

- Similarly, let us take the second and third equations:

$$s + m = s + w$$
$$w = m$$

- We can now see that $w = s = m$, which means that each cup contains the same amount of liquid.
- Let us try a few possibilities.
- If there are six cups in total, then there would be two cups of water, two of soda, and two of milk. Consider the cups of soda in this case. All the cups (six) except two would make $6 - 2 = 4$ cups. But this should be two.
- The same contradictory finding applies to the other cups.
- We can try out a few other possibilities, but the simplest case of all actually applies.
- If there were just three cups, then there would be one cup of each liquid.
- This works out perfectly, since all the cups except two are filled with soda (namely one), all the cups except two are filled water (namely one), and all the cups except two are filled with milk (namely one).

3. *Answer*: 33.33 pounds of potatoes, 16.67 pounds of beets

Solution

Here is the breakdown:

- Let x stand for the number of pounds of potatoes in the mixture.
- Since the actual number of pounds in the mixture was 50, of which x pounds were potatoes, therefore, the number of pounds of beets in the mixture was $(50 - x)$.
- The potatoes cost $0.45 per pound. So, x pounds cost $0.45x$.
- The beets cost $0.30 per pound. So, $(50 - x)$ beets cost $0.30 (50 - x)$.
- The total cost of the potatoes and beets together was, therefore:

$$0.45x + 0.30 (50 - x)$$

- The mixture cost $0.40 per pound. So, 50 pounds cost $0.40 \times 50 = \$20$. This is the cost of the potatoes and beets together:

$$0.45x + 0.30 (50 - x) = 20$$
$$45x + 30 (50 - x) = 2000$$
$$45x + 1500 - 30x = 2000$$
$$15x = 2000 - 1500$$
$$15x = 500$$
$$x = 33.33 \ (= \text{ pounds of potatoes})$$
$$(50 - x) = (50 - 33.33) = 16.67 \ (= \text{ pounds of beets})$$

4. *Answer*: 7.2 ounces

Solution

Here is the breakdown:

- Let x be the number of ounces of oil to be added.
- The original number of ounces in the mixture was 12. So, the total number of ounces in the new mixture will be $(12 + x)$.
- In the original mixture, 40% of the 12 ounces, or $0.4 \times 12 = 4.8$ ounces, was made up of vinegar.
- The new mixture of $(12 + x)$ ounces will have 25% vinegar, or $0.25 (12 + x)$.
- This does not alter the actual amount of vinegar in the new mixture, since we have added oil, not vinegar to it. So, it is still equal to 4.8 ounces, albeit with a different percentage:

$$0.25 (12 + x) = 4.8$$
$$25 (12 + x) = 480$$
$$300 + 25x = 480$$
$$25x = 180$$
$$x = 7.2$$

- This means that 7.2 ounces of oil must be added to the new mixture.

5. *Answer*: 220

Solution

Here is the breakdown:

- Including the square section, the area of the field is $100 \times 80 = 8000$ (square feet).
- The area of the square section is $50 \times 50 = 2500$ (square feet).
- So, the area left for the farmer to plant trees is: $8000 - 2500 = 5500$ (square feet).
- Each tree occupies a $5 \times 5 = 25$ square feet region. Therefore, the number of trees that can be planted is $5500 \div 25 = 220$.

6. *Answer*: 40 minutes

Solution

Here is the breakdown:

- The cold faucet pours water into the tub at 15 liters per minute, and the hot one at 10 liters per minute.
- So, when they are on together, they pour water at 25 liters per minute into the tub.
- The hole jettisons the water at 12 liters per minute.
- The difference between the two is $25 - 12 = 13$ liters per minute, which is what actually pours into the tub.
- The tub holds 520 liters, so the overflow will occur after $520 \div 13 = 40$ minutes.

7. *Answer*: seven transfers

Solution

At the start, the state of the jars can be represented with [8, 0, 0], which says that there are 8 liters in jar A, and 0 liters in jar B and jar C, respectively. Here is a summary of the transfers.

1. Pour 5 liters from A into B, leaving 3 in A:

   $[8, 0, 0] \rightarrow [3, 5, 0]$

2. Pour 3 liters from B into C, leaving 2 in B:

   $[3, 5, 0] \rightarrow [3, 2, 3]$

3. Pour the 3 liters from C into A, which makes 6, since there were 3 liters in A, and leaving 0 in C:

   $[3, 2, 3] \rightarrow [6, 2, 0]$.

4. Pour the 2 liters from B into C, leaving B empty:

   $[6, 2, 0] \rightarrow [6, 0, 2]$

5. Pour 5 liters from A into B, leaving 1 in A:

   $[6, 0, 2] \rightarrow [1, 5, 2]$

6. Pour the 1 liter from A into C, making 3, since 2 were already there, and leaving A empty:

   $[1, 5, 2] \rightarrow [0, 5, 3]$

7. Pour the 3 liters in C into A, which is A's final content. C is now empty, and B retains its 5 liters. The transfer procedure is now complete.

   $[0, 5, 3] \rightarrow [3, 5, 0]$

8. *Answer*: eight transfers

Solution

Let us label the 9-liter jug A and the 4-liter jug B. Both are empty at the start: $[0, 0]$.

1. Fill jug A with water from an external source:

   $[0, 0] \rightarrow [9, 0]$

2. Pour 4 liters into B, leaving 5 in A:

   $[9, 0] \rightarrow [5, 4]$

3. Pour out the water from B, leaving it empty:

   $[5, 4] \rightarrow [5, 0]$

4. Pour 4 liters from A into B, leaving 1 in A:

   $[5, 0] \rightarrow [1, 4]$

5. Empty B:

   $[1, 4] \rightarrow [1, 0]$

6. Pour the liter from A into B, leaving A empty:

   $[1, 0] \rightarrow [0, 1]$

7. Fill A to the top, making 9 liters:

   $[0, 1] \rightarrow [9, 1]$

8. Pour 3 liters from A into B so as to fill it to the top (4 liters), given that there is 1 liter already there, and leaving 6 in A:

[9, 1] → [6, 4]. After eight transfers, A has the required 6 liters.

9. *Answer*: four transfers

Solution

The initial contents of the two jugs can be represented as [0, 0]. When full, their contents are represented by [4, 3].

1. Start by filling the 3-liter jug from an external source:

[0, 0] → [0, 3]

2. Pour the 3 liters into the 4-liter jug. Thus leaves the 3-liter jug empty:

[0, 3] → [3, 0]

3. Refill the 3-liter jug from an external source:

[3, 0] → [3, 3]

4. Fill the 4–liter jug, by pouring water from the 3–liter jug, which has 3 liters in it. This means that 1 liter is poured from that jug into the bigger one, leaving 2 liters in the 3–liter jug, which is the solution.

[3, 3] → [4, 2].

10. *Answer*: 4 hours 43 minutes and 17 seconds

Solution

The breakdown is as follows:

- The first faucet takes two days, or 48 hours, to fill the pool.
- In 1 hour, it fills 1/48 of the pool.
- The second faucet fills 1/72 of the pool in 1 hour, since there are 72 hours in three days.
- The third faucet fills 1/96 of the pool in 1 hour, since there are 96 hours in four days.
- The fourth faucet fills 1/6 of the pool in 1 hour, since it takes 6 hours to do so.
- Together, in 1 hour the faucets will fill:

$$1/48 + 1/72 + 1/96 + 1/6 = 61/288 \text{ of the pool}$$

- To fill the pool, at 61/288 per hour, it will thus take 288/61 = 4 hours 43 minutes and 17 seconds. To see this, start by considering that in 1 hour 61/288 of the pool is filled; in 2 hours, it will be 2 × 61/288 = 122/288; in three hours, it will be 3 × 61/288 = 183/288; in 4 hours, it will be 4 × 61/288 = 244/288; and this leaves 43 minutes and 17 seconds to complete the filling process of 288/288.

Chapter 10

1. *Answer*: 60

Solution

Since each tile covers $5 \times 4 = 20$ square units, then 30 of them will cover an area of $20 \times 30 = 600$ square units. This is also the area of the floor. Since its length is to be 100 units, therefore, its width will be 60: $100 \times 60 = 600$.

2. *Answer*: 7×7

Solution

Here is the breakdown:

- The number of tiles required is 21, and each one must be a square figure.
- So, if we represent a side of a tile with x, then the area covered by each tile is x^2.
- The total area covered by the tiles is, thus, $21x^2$.
- We are told that the area of the floor is 1029 square units. This is the area to be covered exactly by the tiles:

$$21x^2 = 1029$$
$$x^2 = 1029/21 = 49$$
$$x = \pm 7$$

- The negative value has no meaning in this case, so the answer is 7.
- So, each tile is 7×7.

3. *Answer*: 6×12

Solution

Here is the breakdown:

- If we let x be the length of the floor, then its width is $2x$.
- Its area is, thus, $x \times 2x = 2x^2$.
- Now, one tile covers $2 \times 2 = 4$ square units. There are 18 of them. So, overall, the tiles cover an area of $4 \times 18 = 72$ square units.
- This means that the area of the floor will be 72 square units, since the tiles fit perfectly in it. Therefore:

$$2x^2 = 72$$
$$x^2 = 36$$
$$x = \pm 6$$

- We discard the negative value because it has no meaning in this case. So, the length of the floor will be 6 units, and its width, which is twice the length, will be 12 units.
- This means that the dimensions of the floor are to be 6×12.

4. *Answer*: 3, 4, and 5 units

Solution

Here is the breakdown:

- The triangular façade has sides of 5 units each, and a base of 6 units. When a perpendicular is dropped from the vertex to the base, it bisects it, into two units of 3 each (Figure A10.1).

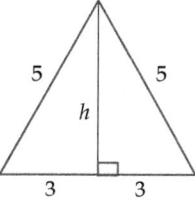

Figure A10.1 Triangular Façade

- This forms two triangular tiles.
- Using the Pythagorean theorem, we can determine the length of the perpendicular, which is also the height, h, of each tile:

$$5^2 = 3^2 + h^2$$
$$h^2 = 5^2 - 3^2$$
$$h^2 = 25 - 9$$
$$h^2 = 16$$
$$h = \pm 4$$

- We discard the negative value. So, the dimensions of each tile are 3, 4, and 5 units.

5. *Answer*: see solution

Solution

The triangle made with the diameter of a circle as base and its two sides meeting at the circumference is a right triangle, as shown in Figure A10.2. If we make two of these, one above and one below the diameter, with all four sides equal, then we will have our internal square carpet to be put into the circular floor, touching it with its vertices (Figure A10.2).

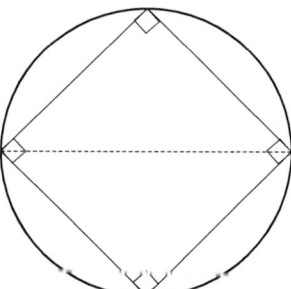

Figure A10.2 Square Carpet in the Circular Floor

6. *Answer*: three L-shaped cuts

Solution

See Figure A10.3.

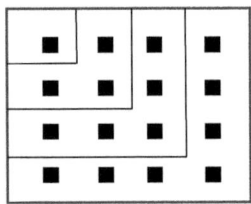

Figure A10.3 Three L-Shaped Cuts

7. *Answer*: 29 feet

Solution

The rectangular floor with the linear path to be designed through it can be sketched as shown in Figure 10A.4. Note that the path could have been drawn from the other corner at the top. Either way, the diagonal would have the same length. On the sketch, the width is marked as 20, the length as 21, and the diagonal as *d* (Figure A10.4).

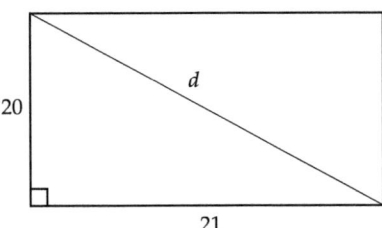

Figure A10.4 Sketch of the Floor

As can be seen, the diagonal is the hypotenuse of a right triangle, which can be easily computed with the Pythagorean theorem:

$$d^2 = 20^2 + 21^2$$
$$d^2 = 400 + 441$$
$$d^2 = 841$$
$$d = 29$$

8. *Answer*: 19.635 square feet

Solution

The diameter of the circle touching the square will equal the length of a side of the square, as can be seen in Figure A10.5.

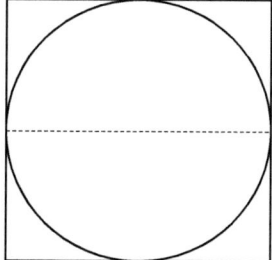

Figure A10.5 Circular Design on a Square floor

Let a side of the square be s. Since the area of the square is given as 25 square feet, therefore:

$$s^2 = 25$$
$$s = 5$$

So, the diameter of the circle will be 5 feet, meaning that its radius will be 2.5 feet. Therefore, the area of the circular design to be drawn on the square floor will be:

$$\pi r^2 = \pi(2.5)^2 = 6.25\pi = \text{ approximately 19.635 square feet}$$

9. *Answer*: see solution

Solution

The partitioning is shown in Figure A10.6. Note that the size of the square does not alter the solution.

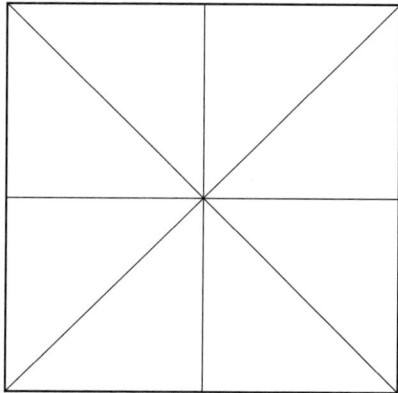

Figure A10.6 Eight Equal Triangles in a Square

10. *Answer*: see solution

Solution

Dudeney's solution is shown in Figure A10.7. Note that the larger square, ACEG, is made up of four equal parts assembled from the original two squares (ABCD and BCFE):

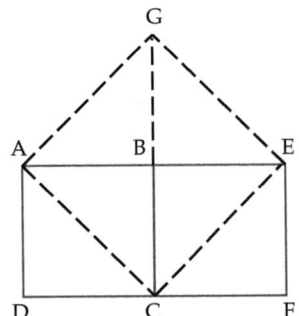

Figure A10.7 Dudeney's Puzzle

Index